Intelligent GIS

Location decisions and strategic planning

Mark Birkin

Lecturer, School of Geography, University of Leeds and Director of Research and Development, GMAP Ltd.

Graham Clarke

Lecturer, School of Geography, University of Leeds.

Martin Clarke

Professor of Geographical Modelling, School of Geography, University of Leeds and Managing Director, GMAP Ltd.

Alan Wilson

Vice Chancellor and Professor of Urban and Regional Geography, University of Leeds.

GeoInformation
International

GeoInformation International
A division of Pearson Professional Ltd
307 Cambridge Science Park
Milton Road
Cambridge
CB4 4ZD
and associated Companies throughout the world.

Copublished in the Americas with John Wiley & Sons Inc,
605 Third Avenue, New York, NY 10158

First published 1996

British Library Cataloguing in Publication data
A catalogue entry for this title is available from the British Library.

ISBN 1-899761-25-X

Library of Congress Cataloguing in Publication data
A catalogue entry for this title is available from the Library of Congress.

ISBN 0-470-23614-0

Printed in the United Kingdom
by Bell and Bain, Glasgow
Set in 11/12.5 Palatino

CONTENTS

ACKNOWLEDGEMENTS

The authors would like to thank Vanessa Lawrence of GeoInformation International for her continued enthusiasm and support for this project, despite the passing of many deadlines! Thanks also to Sarah Langman Scott for managing the production of the book through its final stages. Much of the material in this book has been based on projects undertaken by GMAP Ltd, and we gratefully thank both the staff of GMAP and its clients for their support. Thanks also to Alison Manson and 'AJ' for their contribution on the artwork of the book. Alison and AJ work at the graphics unit, School of Geography, University of Leeds.

We are grateful to the following for permission to reproduce copyright material:

The publishers, GIS World, Inc., for Figure 3.3; Toyota (GB) Ltd, for Figure 4.1; UCL for Figure 4.2; Professor Les Worrall, for Figures 4.6 and 7.2; Marks & Spencer plc, for Figure 4.8; Routledge, for Figure 8.1.

While every effort has been made to trace the owners of copyright material, in a few cases this has proved impossible and we take the opportunity to offer our apologies to any copyright holders whose rights we may have unwittingly infringed.

LIST OF FIGURES AND PLATES

LIST OF TABLES

CHAPTER 1

Introduction

1.1 AIMS

The principal aim of this book is to present an account of how it is possible to integrate geographical information systems (GIS) technology with geographical modelling techniques in order to help contribute to better understanding and better planning within organizations in both the public and private sectors. The account is based mostly on the authors' collective experience of working in this area with large organizations in both these sectors. Our focus is partly on the theory of how this integration can be achieved, but mainly on the extensive use of practical illustrations of integration from our own work. Much of this has been undertaken through GMAP Ltd, a private limited company owned by the University of Leeds. All four authors were instrumental in different ways in both establishing and then developing GMAP. One author (Martin Clarke) is currently the managing director and another (Mark Birkin) the director of research and development. We believe that this focus on practical illustration is the only valid way to make the point about the value and relevance of integrating two different although related sub-disciplines within geography.

Our title, *Intelligent GIS*, might cause a few raised eyebrows. By implication, it may be thought that we feel that all existing GIS applications are 'unintelligent' in some way. This is not our intention. What we wish to demonstrate is that through the integration of conventional GIS software and model-based analysis we get something that is bigger than the sum of the parts. This 'something' is improved knowledge about geographical systems (such as the retail market for cars or the provision of health-care facilities in a region) and accurate insights into the impacts of potential changes within those systems. When assimilated and used sensibly, this analysis can result in levels of increased 'intelligence' about that system. This intelligence should lead to better planning and decision making than if this capability did not exist. Our use of the word 'intelligent' is therefore closer to that in 'military intelligence' than in 'artificial intelligence', and closest in the literature to the work on 'spatial decision support systems'. Our intelligent GIS framework is sketched out in Figure 1.1. Thus, we argue that the intelligence comes from two directions: first, the use of appropriate statistical or mathematical modelling methods, and second, through more useful and informative model outputs in the guise of performance indicators. In

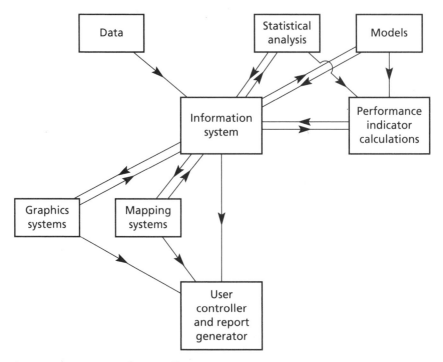

Figure 1.1 Structure of an intelligent GIS

particular, however, we shall argue that it is only through the use of geographical modelling methods that the key components of market analysis can be assembled. These arguments are developed in some detail in Chapter 3.

A second aim is to demonstrate the relevance of geographical analysis in providing a critical input to strategic and operational planning in most organizations. Organizations, whether public or private sector, supply goods or services to meet identified demands. The demand for goods or services is not uniform across geographical space, whether the level of focus is between different countries, between regions of a country, or between census wards of a city. Different types of people with different tastes and levels of income live in different locations at varying densities. Quantifying this variation in demand is an important component of any planning process. To meet this demand, suppliers usually establish a distribution channel for their products. Some of these channels involve locating outlets (shops, petrol stations, hospitals, etc.) in distinct physical locations as part of a supply network. Other channels, such as mail order, telephone banking and so on, have more flexible geographical characteristics. However, in all cases there is an interaction between demand and supply that involves the flow of something, whether this be money, information, people, or goods. It is an understanding of the complex inter-relationship between demand and supply and the way this varies over space and time that is at the heart of many organizations' management responsibility. By being able to plan effectively on the basis of this understanding, organizations should be able to improve their performance in the market place. The supply-demand inter-relationship forms the substance of many of the subsequent chapters on applications of intelligent GIS.

In attempting to demonstrate the relevance of applied geography, we are targeting three broad types of reader. First we wish to persuade the academic geographer that there is true value and relevance in his or her subject in the commercial world. (There is also substantial value in it outside the commercial world, but this has been the subject of many texts: see especially Gould (1985).) Geographers have not traditionally been particularly good at 'selling' the relevance of their analysis to the outside world. Economists and management scientists have been much better at this, and hence often get taken more seriously than geographers in business and commercial circles. We hope this book might shift this balance in some small way.

The second type of reader is the chair, managing director, or market analyst within a private or public sector organization. For most of these individuals, formal geographical education probably ended at the age of 16. Their hazy recollection of geography as a school discipline will most probably consist of a concern with lists of place names associated with certain features, oxbow lakes, and an obsession with sheep farming in Australia! Typically they will have little knowledge of what contemporary applied geography can bring to their organization. Again, we wish to change their view in some small or possibly bigger way. Overall, therefore, we hope to demonstrate that not only is geography relevant and important, it can also be exciting in that it can sit comfortably within the heart of the business processes that large organizations undertake.

The third type of reader we wish to reach is the undergraduate student. These students may have become accustomed to thinking that quantitative geography is dead and buried or, if it still lives, is represented by those rather boring statistical practicals that they never understand or will never use again. We hope this book will illustrate the applied relevance of GIS and spatial models and their growing importance in the real world. We have made a conscious effort to make the results of our work accessible to those readers who have not got a detailed technical understanding of either modelling or GIS. References to more detailed material are provided at the appropriate points in the subsequent chapters. However, at the outset, for those interested in an overview of mathematical modelling in geography and planning, Wilson and Bennett (1985) would serve as a useful guide. There is now a plethora of texts on GIS, but the two-volume collection of papers edited by Maguire et al (1991) probably remains the seminal review of work in this area. We hope that our policy of keeping the technical detail away from the reader allows undergraduates in geography and related disciplines to focus on the main aims of the book as described above. It is through raising students' interest and excitement in how their discipline can be used in a relevant way that they may become more willing to get to grips with some of the technical detail that is a prerequisite for them to begin contributing to applied spatial analysis.

A third aim is to describe the wide range of applications of 'intelligent GIS' that have been developed across the spectrum of market sectors in the economy. It is easy to forget that the classic location theories developed within geography over the last 150 years were undertaken by geographers who had real concern with practical problem solving. Von Thunen was a farm estate manager interested in improving the efficiency of production and distribution. Alfred Weber, the industrial location analyst, was concerned with the optimal location of industrial plants given fixed supply and market locations. Walter Christaller's concern with settlement location was inspired by his fascination with the distribution of villages, towns, and cities in southern Germany. We have argued that much of the research that underpins the geographical models presented in subsequent chapters of this work was motivated by similar applied problem-solving objectives. Indeed, the development of the family of spatial interaction models that has been the core of our research has resulted in a contemporary rewriting of the theories of agricultural, industrial, residential, and service location (Birkin and Wilson 1986, Wilson and Birkin 1987, Clarke and Wilson 1985a).

It is sometimes useful to classify research into three distinct types. Basic research is concerned with the quest for knowledge, often through hypothesis testing. Applied research uses the results of basic research in particular contexts. Strategic research uses research findings in guiding the direction that organizations pursue. A good example of these three types of research is found in the pharmaceutical industry. Biochemists and genetic engineers, who are engaged in basic research, might study how a particular molecular compound affects aspects of human DNA. Applied research will examine whether this process can be used effectively to treat a disease or medical condition. Strategic research will examine the possibility of turning this into a commercial drug to be marketed in specific geographical territories.

Demonstrating the importance of strategic research in focusing the basic research agenda is a fourth aim of this book. Some academics believe that the research process is a linear one, from basic research to applied research and, eventually, through some technology transfer process, filtering into commercial application. Traditionally, the university sector has focused on the first two parts of this process. There has been no great incentive to partake in the last part within the university environment, although a few entrepreneurial academics have done so outside it (see Bryan 1995, Clarke et al 1995).

Over the last decade or so this traditional model has begun to collapse. We have argued elsewhere (Clarke 1990) that the research process should be (and in some cases is) seen as a cyclical process between pure or basic and applied. The fundamental difference here is that the endeavours and frustrations of applied and strategic research feed back in defining and redefining the basic research process. Academics can spend at least part of the research effort focusing on improving methods and tools that are directed at problems of relevance to industry, commerce, and government. This is not to exclude 'blue skies' research from the agenda or to suggest that the only relevant research is that which is driven by short-term problem solving. However, the recent UK White Paper on Science and Technology argues strongly for developing links between producers and users of research, and that university research itself should be more focused on improving the UK's international industrial competitiveness. The benefit of this focus is that work in this collaborative environment can be exciting – working with organizations that appreciate the benefits of research outputs can be intellectually demanding and genuinely rewarding. Universities are also recognizing that the technology transfer process is a legitimate academic activity that should be seen alongside traditional research activity as within the portfolio of desirable endeavour. Indeed, many universities have set up their own technology transfer companies (of which GMAP is but one example) specifically to exploit the links between academia and commerce.

It may be useful to give an example of how this cyclical research can work. In Chapter 9 we describe how a method to locate a network of auto-dealers optimally in a region has been applied. The method was first developed as a research tool to examine urban spatial structure in the 1980s. With some modifications we successfully developed it for use in the auto industry through work with a GMAP client. However, because of the computer-intensive nature of the model it was only possible to apply the approach to a single region of the country at a time. It was estimated that to run a UK optimal scenario would take around three months on a SUN workstation. It was clear that this was unacceptable, and we had to search for an alternative solution. At the time, discussions of collaborative academic research had started with colleagues in the computer studies department at the university. This particular problem was raised, and we eventually applied for and received a research grant from the Department of Trade and Industry (DTI), along with the University of Edinburgh Parallel Computing Centre, to explore the use of massively parallel computers to address this problem. Twelve months of university research resulted in the reformulation of the optimization problem and in the execution of the national scenario in minutes, not months. A full discussion of the research can be found in Birkin et al (1995).

1.2 GEOGRAPHY AND BUSINESS

In spite of the intuitive nature of the model which shows that people are increasingly less likely to visit retail or service centres which are of greater distance from their residence or workplace, it nevertheless comes as a surprise to many on the marketing side of organizations that, rather than market share being evenly spread across a region, high levels of penetration are concentrated around branches. Conversely, there are whole tracts of residential areas where some players fail to build any significant share of the numbers of customer relationships because of gaps in their supply network, despite having competitive products. In our experience, there is a widespread misconception that market share is fundamentally dependent on product pricing and other complicating factors such as brand awareness and advertising. The truth is really far more complex. Whereas we would not deny that, at a national level, brand strength, product competitiveness, and advertising spend all play a part in determining the market share of

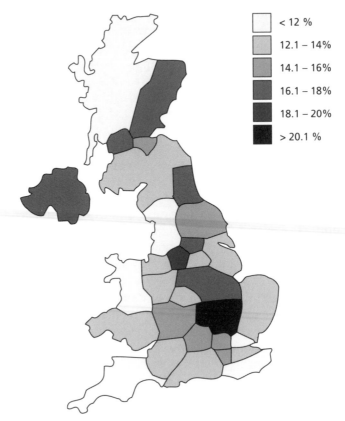

Figure 1.2 Market share for a major motor vehicle retailer at the regional level
(*Source*: GMAP 1994)

the business, detailed analysis at the local level demonstrates that market share is significantly determined by the location of outlets with respect to local market demand and to competitor locations. In fact, even for major clearing banks where one might consider branch-based supply to be more or less ubiquitous given the density of their networks and their presence in almost all towns of significant size, we have shown that market share at the local level can vary from over 50 per cent to less than 5 per cent between residential areas within the same region.

As an example, let us consider a large organization such as a motor vehicle retailer with a share of the UK market of 16 per cent. Figure 1.2 shows at a regional level that this varies from less than 12 per cent in many areas to over 20 per cent in others. Figure 1.3 shows that in one particular region (in this case Bedfordshire and Hertfordshire) it varies from less than 8 per cent to over 33 per cent! Typically it would be rare for such an organization to consider anything other than its share of the national market. Certainly, few organizations look at anything below TV region level, and yet, as we will attempt to show throughout this book, it is at the local level, the space in which potential customers live and work, that the battle for market share needs to be fought.

We argue that this implies that companies need to rethink the way in which they look at the market. First, rather than thinking of their business nationally or even globally, the real trick is to focus on much smaller markets which correspond more closely with the actual space in which existing and potential customers live, work, shop, open accounts, and make transactions. This implies analysis at the local level, which in turn implies devolved, decentralized management empowered to act at that level. Secondly, there has been an unhealthy, narrow focus on the performance of individual branches. Our experience is littered with examples of companies that believe that the performance of branches is mainly determined by the personality and skill level of branch managers, whilst ignoring the level of demand in the local market and the strength of localized competitor branches. We suspect that there may well be towns in Britain (and in other countries) where every retail and service company believes its local manager is underperforming, whereas, in truth, all are competing in an oversupplied, saturated market which simply does not have enough business to go round. In equal measure, there are likely to be retail centres where the exact opposite is true. Thirdly, if an empowered local management really are to have positive influence in improving the performance of the business at the local level, then they need to be in possession of accurate management information on localized performance patterns within the region under their jurisdiction, and to have appropriate decision support tools which help them manipulate the local market to the benefit of the organization and the organization's customers.

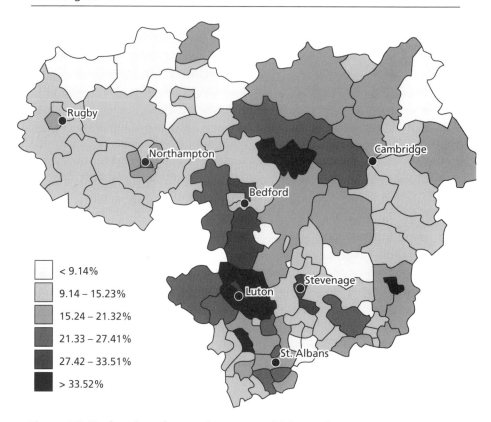

Figure 1.3 Market share for a major motor vehicle retailer at postal district level (*Source*: GMAP 1994)

We believe the evidence of the importance of local business geography is overwhelming. We hope the reader will also be convinced of this by the end of the book!

1.3 SPATIAL MODELLING

To appreciate what contemporary spatial modelling has to offer, it is useful to have a little background on how the subject area has developed over the last 30 or so years. This not only gives a historical summary but is also important in illustrating that spatial analysis covers a wide range of approaches, and that its popularity within the discipline of geography has waxed and waned considerably during the period in question.

The first phase we can identify may be loosely described as 'statistical description'. In the early 1960s, geographers discovered that many simple statistical methods had important applications in spatial pattern analysis.

These techniques, such as point pattern analysis, nearest neighbour analysis, correlation, and regression, quickly established themselves in the geographer's toolbag of methods, notably following the publication in 1965 of Haggett's *Locational Analysis in Human Geography*. As the decade progressed the tools became more refined, with multivariate methods, such as factor analysis, emerging. Although it is often pointed out, it is worth repeating that the so-called 'quantitative revolution' was always more of a statistical than a mathematical one. Much of this statistical work continues today, particularly through the methods associated with discrete choice theory and categorical data analysis (for example, Wilson and Bennett 1985, Wrigley 1985, Haining 1993).

A second phase ran parallel with the development of statistical methods in geography. This saw diffusion of methods from North America to the United Kingdom in the 1960s, in particular land-use and transportation models. Hence, this second phase was largely associated with mathematical models and their application. This work was not focused centrally within geography or indeed within universities. The majority of research was undertaken in a variety of central and local government departments or in quasi-government organizations such as the Centre for Environmental Studies. As a consequence, much town planning in Britain for a period became quantitative and often model-based; this was most clearly expressed in the structure plans prepared by local authorities (see the review of Batty 1989, Bertuglia et al 1994b). A significant conceptual step in this period was the derivation of a wide family of spatial interaction models which used entropy-maximizing methods (Wilson 1967, 1970). This gave the 'gravity model' (already of considerable interest to geographers) a clearer theoretical foundation, and enabled mathematical methods to be established within human geography alongside statistical techniques. Much applied work took place with the development, calibration, and application of land-use transportation models, in particular with respect to a variety of city-regions around the world. This work was characterized by conditional forecasting, where the 'conditions' were representations of the planning policies being explored. Furthermore, attempts to construct comprehensive models took place. This was driven by the realization, first formally recognized by Lowry (1964), of the major interdependencies between sectors of the urban system and the consequent need to represent them in modelling. Although this style of research fell out of fashion (for reasons discussed below), it laid the foundations for a decade or more of theoretical developments on the dynamics of urban spatial structures at the School of Geography, University of Leeds (see the summary paper of Clarke and Wilson 1985a). This, in turn, provided a much clearer understanding of model structures and dynamics and an increased level of confidence to move forward into a new era of applied research.

The third phase, which may be called 'quantitative backlash', can be identified during the 1970s. A number of influences changed the perception of urban modelling and its contribution to planning. First there was the liberal planners' view that the relationship of model-based research to planning was less than satisfactory. The models had failed to live up to the claims that their proponents had made. On the one hand, the problems of urban areas were getting worse and, on the other, the models could demonstrably be shown not to work. This type of criticism is epitomized in Lee's (1973) paper 'Requiem for large scale models'. The second influence came from the advent of radical urban geography, dating from Harvey's *Social Justice and the City*, also published in 1973. It was argued that deeper structures in society determined urban change and that these had been ignored by modellers. Both liberal and radical critics commented upon the static nature of existing urban models, and in some cases methods were criticized on this ground alone (see also Sayer 1976, 1979).

A broader element of the radical critique was that the theory underpinning modelling was simply wrong. Put at its strongest (and sometimes most naive), there was an associated implication that anything 'mathematical' was 'scientific', and therefore 'positivist', and therefore wrong. More productively, it can be argued that urban modellers failed to build on theories related to the main agents of urban structure and development, and relied too much on crude, neo-classical, economic foundations. The first part of this latter criticism certainly has substance; the second was often applied to models for which it was not true. It is better simply to agree that modelling offers a set of tools for handling complexity, and that the skills of the modeller are likely to be relevant whatever the theoretical base.

However, the effect of the criticisms directed towards spatial modelling was to cast the field into a dark age, where developments in model-based methods in the early 1980s were undertaken by a small group of committed researchers (including these authors). In the middle of the 1980s we witnessed the first signs of the emerging GIS world, and many former spatial modellers became engaged in GIS research. We noted at an early stage (Birkin et al 1987) that there was a natural complementarity between GIS and spatial modelling. We can also note a number of trends in recent years that encouraged us to believe in a future for modelling. These include the following:

1 The development of a range of new methods including microsimulation (Clarke and Holm 1987), methods based on artificial intelligence concepts such as genetic algorithms and neural networks (Openshaw 1992), fractal models (Batty and Longley 1994), and 'competing destinations' and other interaction hybrids (Fotheringham 1983, 1986). For reasons we shall describe later, better experience in the use of existing methods has also assisted in their wider application.

2 A much greater availability of data that form the necessary inputs into spatial models. Interestingly, massive amounts of spatially referenced data are being automatically captured by private sector organizations such as banks and retailers through point-of-sale systems and Automated Teller Machines (ATMs). To access these data it is almost always necessary to work with a commercial partner. At the same time, there has been a growing realization that data are an important corporate asset, and new techniques have been encouraged to capture data more routinely.

3 A greater interest in spatial analysis from a range of commercial organizations. This interest was first triggered by the development of geodemographic segmentation tools such as A Classification of Residential Neighbourhoods (ACORN), but this has moved on into a wider interest in the types of predictive tool that are described in this book (see also Longley and Clarke 1995).

4 The availability of powerful desktop computers and related software, including GIS. Ten years ago computer models still had to run on mainframes. Since then there has been a migration to PCs and workstations, making modelling tools both more accessible and transferable. The availability of powerful relational databases and mapping software has allowed better representation and interrogation of model results.

These trends and other factors have seen a renewed interest in the possibilities for planning using computer models. Interestingly, the one area where enthusiasm is still weak is in urban planning, for reasons we shall address in Chapter 7. It is probably true that applying spatial models in retailing, health-care, or the auto industry is less challenging than in urban planning (though far from trivial). The original motivation of spatial analysts 30 years ago was to produce models that would lead to better city planning. If this has not yet occurred, then we can take some comfort in the fact that there have been substantial spin-off benefits in sub-systems such as retailing and health-care.

1.4 CONTENTS OF THIS BOOK

The rest of this book is organized as follows. In Chapter 2 we examine the contribution of locational analysis to the planning process undertaken in most organizations. We examine the different elements of planning – objective setting, analysis, and design. For many organizations there will be important geographical considerations in each of these elements. For example, a retailer will need to set market-share objectives by country and region, and to undertake analysis that explains why performance varies by outlet, by region, and so on. The retailer will then have to design solutions to help meet objectives on the basis of the analysis undertaken. Should they open new stores? Where should these be located? What return on investment might these generate? We also describe in this chapter some of the organizational issues that need to be confronted in this planning process. Who in the organization is responsible for getting to grips with the type of geographical analysis described in this book? As we shall demonstrate, it is an area that often falls between different spheres of responsibility – marketing, planning, property, and so on. The importance of getting the organizational issues right in this area cannot be overestimated – a theme we shall revisit in subsequent chapters.

Chapter 3 explores the potential for GIS to contribute to the solution of business problems in both the public and private sectors. The major point to make is that for GIS to succeed it needs to be integrated with powerful spatial modelling capabilities to provide true decision support functionality. In this context we review the development of GIS in recent times, explore the mainstream components of a typical GIS, and describe some of the more interesting applications. We then develop the theme of linking spatial analysis with GIS by reviewing the main types of capability that spatial modelling can provide. We also examine the recent debate about making GIS more analytical, and conclude by stating that ultimately it is modelling that will provide the real business benefits to industry and commerce, but that the packaging of models within a GIS framework will improve the usability and user-friendliness of the models.

In the same way as Chapter 3 provides a critical review of the role of GIS in business and service planning, Chapter 4 reviews the variety of approaches to locational analysis traditionally seen in the retailing and marketing literature. Each of the major approaches is discussed in terms of its strengths and weaknesses. The major conclusion, as in Chapter 3, is that the levels of complexity that are necessary to handle real-world impact assessments must be addressed through robust modelling methods rather than weak analogue or catchment-area methodologies.

The next six chapters are all concerned with the application of what we term 'intelligent GIS' in different business environments. We use a number of different examples to illustrate the argument, all based on work that GMAP has undertaken with its clients. Chapter 5 demonstrates how an effective analytical capability can be developed. Retailing is a fascinating example in many ways, partly because we all have frequent experience of using retail facilities as consumers and probably have our own opinions on what is good and bad. More importantly, geographers have long been interested in retailing: Christaller's central place theory was largely concerned with the level of retail service provision in different locations, and one of the first spatial interaction models to be developed specifically addressed the problem of defining retail trade areas (Huff 1963). Retailing is also a significant sector of the economy which invests heavily in new capital developments. Being able to demonstrate that these investments will generate acceptable levels of return is an important requirement for retailers.

Health-care is another topical subject that has rarely been out of the media headlines in the UK and USA over the last few years. In Chapter 6 we explore how intelligent GIS can contribute to better planning and analysis in health-care systems. Again, geographers have long been interested in many aspects of health-care. Often the product of this interest has appeared under a 'medical geography' label, examining such issues as the link between health status and deprivation, the spatial transmission of disease, and the relationship between access to health-care and its utilization. Much of this work has been descriptive, exploring relationships between health-care-related variables. We aim to show in this chapter that it is possible to build upon this analysis and move towards predictive and prescriptive work through the use of forecasting methods. In particular, we focus on how methods can be developed that allow planners involved in the new purchaser–provider 'internal market' to make better-informed decisions. In our work in health-care planning over the last decade, we have been aware of the need to take the emotion out of planning in the National Health Service (NHS). Despite its being a data-rich environment, it has come as a surprise how often serious analysis of health-care planning issues is avoided. The recent Tomlinson Report on the future of health-care provision in London (HMSO 1992) was notable for its almost complete lack of geographical analysis, yet the impacts of the Report's recommendations will have profound implications for residents living in certain parts of the capital. Without access to appropriate planning tools, health-care planning decisions will be more imperfect than they might be. In a retailing context the only losers are the shareholders: in health-care the implications of poor planning are much more widely felt.

Urban planning shares many of the characteristics of health-care planning – it is complex and difficult and there are real questions of social costs and benefits to be addressed. In Chapter 7 we attempt to illustrate, through a range of applications, how contemporary applied analysis can throw light on the nature of urban planning issues and help to guide the policy-making process. We note that the practice of urban planning has undergone a series of metamorphoses over the last 20 years. In the 1970s, for example, there were a number of attempts to use integrated urban models as inputs into the structure-planning process in British cities. This type of analysis has now largely disappeared. Town planning is more of an enabling than a proactive process. However, urban planning takes place under other headings – education, training, transport, and housing departments have important inputs into the urban development process. GIS has also been embraced by most local authorities, usually as an operational tool but occasionally in more strategic research areas. We examine some of these applications in Chapter 7.

In Chapter 8 the focus of attention switches to water planning in its broadest sense. Once again, we argue that there is an explicitly geographical focus to understanding the demand for and supply of water. Although we argue in Chapter 3 that GIS has probably had a greater track success in the environmental field, there are still good arguments for improving the level of spatial analysis by introducing models. We address this issue in relation to the supply of water, the measurement of water quality, and new research aimed at estimating small area variations in water demand.

Chapter 9 focuses on the use of intelligent GIS in the automobile industry. The importance of this industry to Western economies cannot be overestimated in terms of both gross domestic product (GDP) and employment. Individuals with access to a car have enhanced levels of mobility and freedom, but generated along with these are the related problems of pollution and environmental degradation. The retailing of cars is an interesting process, with manufacturers controlling supply through the appointment of exclusive franchise dealers, which in the main sell a single manufacturer's product. The manufacturers spend a good deal of time and effort planning their development strategies and dealer networks, and as we shall see this offers significant opportunity for the use of intelligent GIS.

Our final case study (Chapter 10) is taken from the financial service market. This market has undergone dramatic transformation since the mid-1980s and now operates in a far more competitive environment. As well as searching for new business in new markets, many financial organizations now believe that they have too many high street branches, and are looking to a period of rationalization. Despite the extensive use of geodemographics, often linked to GIS packages (see Chapter 3), there is still an urgent need to increase the predictive capabilities of current branch turnover estimation techniques. After a broad review of the importance of geography in these markets, we illustrate the argument with a case study drawn from Australia.

Some concluding comments are offered in Chapter 11. In particular, we focus on the business benefits that can accrue when intelligent GIS is undertaken.

1.5 COMPLEMENTARY READING

It is important to recognize at the outset that there is an increasing awareness within the GIS community that greater analytical power is often needed to solve complex, real-world location problems. Since the mid-1980s there has been a large volume of papers and research conferences addressing the issue of 'spatial analysis and GIS'. This has reached an advanced stage in texts such as Fotheringham and Rogerson (1994) and a number of recent research papers (Fotheringham 1994, Fotheringham and Xie 1995, Batty and

Xie 1994a, b). This research is particularly important as it begins to address the questions not only what spatial analysis can do for GIS but also of what GIS can do for spatial analysis. Whilst recognizing this importance, we still believe much of the research is driven by academic interests in solving difficult statistical problems associated with the handling of spatial data. This is fitting for academic geography, but does little to solve our basic concern about the appropriateness of GIS for business users (the latter term is used to refer to all those concerned with planning services in the real world). It is this gap in the market which this book aims to fill.

We also recognize that the literature relating to GIS and business is growing too. Both Castle (1993) and Grimshaw (1994) provide excellent introductions to GIS for the business community, with a strong emphasis in Grimshaw's text on getting GIS introduced into the company. We strongly recommend readers new to GIS to have a look at these complementary texts: the drawback to them is that they pay only lip-service to adding greater spatial analytical power to GIS. Finally, Longley and Clarke (1995) provides a set of 'state-of-the-art' essays on GIS for business and service planning, focusing on issues of data, geodemographics, and customized versus proprietary GIS use.

CHAPTER 2

Location and planning

2.1 INTRODUCTION: THE ELEMENTS OF PLANNING

Planning is an intrinsic activity in any kind of organization. In this chapter, the concept of planning is unpacked so that its main components can be identified. It is argued in this introduction that there are three main elements to planning, namely objective setting, analysis, and option appraisal, and each of these is explored in turn in the following sections (2.2, 2.3, and 2.4) for a variety of types of organization. In Section 2.5, the ways in which these elements are integrated into management structures – or not integrated – are explored. The construction and implementation of plans are the subjects of Section 2.6 and some concluding comments are offered in Section 2.7.

Any organization, implicitly or, increasingly, explicitly, will have well-defined *objectives*. The first element of planning must be the articulation of these objectives, which will underpin the organization's policy or strategy. In order to make any progress at all, it is necessary to have an understanding of the organization itself, and how the organization functions within its environment; and we can characterize this component of planning as *analysis*. The aim of any planning activity will be to implement some kind of action to achieve objectives, and the third element of planning to be identified at the outset is the discovery of *options* – essentially a process of *design*, though the concept is used here much more broadly than with its narrower, colloquial connotations.

These three activities involve very different kinds of thought process, and this is one of the reasons why planning is often difficult: the highest levels of skill in each are not typically found in the same person, and it is not easy to assemble planning teams within which these skills can be integrated effectively. This task of team building and integration is made more complicated by the fact that the main dimensions of management structures do not usually facilitate it; it is this issue which is explored in Section 2.5.

As the argument develops in subsequent chapters of the book, the planning tasks of a variety of organizations will be addressed. How best can we characterize organizations for this purpose? There are various possible dichotomies, of which market/non-market and public/private are probably the most useful – the former because they are often inherently different in the way they relate to customers, the latter because they differ in their management structures. However, it has been one of the features of the 1980s and 1990s that these distinctions are breaking down, and we need to take that into account in the analysis which follows. The examples which are explored in the rest of the book are concerned with:

- retailers;
- health-care;
- local government, and particularly urban planning;
- the water industry;
- the motor industry;
- financial services.

These can be used to illustrate the specific focus of the book within the broader field of planning: that is, the *geographical* component. It will be argued that geography is both critical and often neglected or underplayed, and that a crucial element of planning, which will integrate the three outlined above, is the *intelligent* geographical information system (IGIS).

Retailers and planners of financial services have a direct interest in the location of both the customers and the outlets which serve these customers. The motor industry is a special case of retailing from this perspective. Health-care managers should have an interest in providing adequate access to services for their patients. Local government, through urban planning, is concerned with the development of spatial structures which integrate many issues of location and access for their populations (and indeed in the provision of their own services). The water industry is, again, concerned with delivery to customers in relation to the sources of supply. We will show that by getting the geographical components right – in relation to objectives, analysis, and the development of options – there are tremendous gains to be made in the effectiveness of planning for all concerned with the outcomes of such processes – organizations and customers.

2.2 OBJECTIVES AND ASSOCIATED STRATEGIES

The objectives of the organization will be determined in the first instance through its (conscious) positioning. This will be represented by its portfolio of products and/or activities as well as the markets and income flows which support the processes that generate these. This specification will be considered in more detail in the next section, and for present purposes we will take it as given.

Objectives will then be stated in terms of efficiency and effectiveness and can usually be summarized under three headings:

- quality and standards;
- market share;
- financial indicators.

Quality and standards will be determined according to the markets or populations to be served. These will be subdivided in various ways; for example, into market niches for different kinds of car, or into population groups with different kinds of health-care need. Most organizations will have an interest in market share. We will show in the following chapters that this concept (and that of market penetration) is particularly valuable at fine geographical scales. Financial objectives can then be stated in relation to the type of organization and its policies. These may range from profit maximization for a private sector firm to balancing the budget for the managers of a public facility. All kinds of organization are likely to have some financial indicators in common, such as the returns to be achieved on capital.

This kind of portrait of the organization can be drawn up for the current situation. Its objectives can then be related to various kinds of improvement (in quality, say, or cost reduction); in maintaining the status quo in market share, possibly defensively as competition is perceived to increase; in seeking to grow; in seeking to minimize risk; in wanting to achieve geographical expansion; and so on. As objectives are articulated more explicitly, targets for achievement can be represented formally in terms of *performance indicators*, a generic term which encapsulates all the kinds of information discussed informally in this section. In order to make progress with this account and to be able to generate the performance indicators, we must have an effective analytical capability, and this is the issue to be pursued in the next section.

2.3 ANALYSIS

The organization and its environment must be understood in depth. This will involve devising a suitable classification of products and activities, in relation both to all aspects of their production functions and to markets or populations. The products and activities so defined, and the processes of sourcing inputs, production, distribution, and delivery (and the capital infrastructure which supports these), will have to be articulated in detail and mapped on to a picture of the structure of the organization, its main units, and the associated costs and income flows.

The same analysis must be conducted for all relevant organizations within the environment of the organization – especially in relation to competitors – and, of course, for the environment itself. Small-area demography, particularly, will be the basis of markets and needs analyses, economic forecasting, and geographical forecasting in relation to shifting spatial patterns. In other words, a comprehensive picture must be drawn, though it is very rare for this to be achieved in practice. All of this information can be summarized in charts and maps.

A conventional way of building on this analysis and moving in the direction of the development of strategies is to conduct a strengths, weaknesses, opportunities, and threats (SWOT) analysis: strengths and weaknesses relate to the organization itself, opportunities and threats to its environment. The analysis of strengths and weaknesses will include a review of market share which has been achieved (or quantity of service which has been delivered), and a critical part of the argument which follows in the examples in the rest of this book is that, ideally, this should be done for very small geographical areas. Typically, market share is known for large areas, such as a country or a television region; but the studies we describe later show that when this is broken down geographically, such market shares, say within a large region of a country, can vary by a considerable factor. Thus a performance which looks to be 'average', and perhaps therefore satisfactory, will be seen to be a mixture of the good and the bad. Such analyses obviously form the basis for identifying new opportunities. They will also suggest targets: what has to be done to achieve more uniformly the best market shares which have been identified on this basis?

The analysis of opportunities and threats has been partly addressed through such indicators as market share, as mentioned above. It will be necessary to move towards a more detailed understanding of the variations in market share by drawing up inventories of competitors, for example. It will then be possible to identify opportunities more systematically: a low, small-area market share may be due as much to the existence of high levels of competitor activity as to poor organizational distribution. There will be equivalent questions for public sector service providers. The argument will become quite complicated when niche markets are taken into account in particular product areas – and these are likely to be fragmenting and multiplying in contemporary conditions.

This argument can be summarized by saying that what is needed is an integrated and detailed management information system based on key performance indicators (which can be directly related to objectives) for both the organization *and* its competitors. To see what this would involve, let us consider first a retailer.

Assume that the retailer is selling a well-defined range of products through a network of stores and that people have to be attracted into the stores to effect their purchases. The competitors will obviously be any similar stores – say bookshops – but also any other kinds of store which also sell some or all of the same range; for example, department stores or markets. Where individuals choose to shop will depend on a complex set of individual circumstances. It is certainly not true, for instance, that in general they go to the nearest store. Thus, to have an effective base for the analysis, it will be necessary to represent all relevant retail activity in quite a large region. This can be considered to be made up of a large number of residential areas or zones, and of a set of shopping centres and the stores within these. The inventories of both markets and stores can be established. The basic data which must underpin any good analysis are then contained in the *interaction matrix*, which links people by zone of residence with shopping centres. (We leave aside for the moment the complication of people shopping from a work base, which can be straightforwardly added.) It is the lack of this matrix and the understanding of its importance which is the critical weakness in many analyses which are commonly carried out.

This matrix is potentially quite large: if there are 100 residential zones, 20 shopping centres, 10 types of people and 5 products for the store type, then there are $100 \times 20 \times 10 \times 5 = 100\ 000$ elements; and these numbers are quite modest in relation to what is often needed for a realistic representation.

This means that the information system must be computerized. Once this base is established, it is possible to compute all the relevant performance indicators. The information system can then be constructed in such a way that it can be interrogated by a user, and utilized as the basis for SWOT analyses and the development of strategies.

There is one further advance to be noted, which converts what is essentially a sophisticated database into an intelligent information system. It is that the system can be modelled: the whole working system can be represented as a mathematical computer model which explains and replicates the flows which represent the heart of the system. It is possible to do this in spite of the idiosyncrasies and complexities of individual behaviour, because the predictions can be made for groups of people on an averaging basis.

This step is critical, because it becomes possible to use the system for 'what if?' forecasting. For example, it would be possible to 'open' three new stores in the region and to rerun the model. This would generate predictions, for instance, of the revenue which would be attracted to each of the new stores (and indeed the losses from any existing stores in the group, and those of competitors). The strategist, therefore, can conduct experiments on the computer and seek to work towards an optimum store network for the organization in the region. It means that, on the one hand, expensive mistakes can be avoided; and, on the other, profitable new opportunities can be identified and the investment made with some confidence.

The argument can now be summarized (and, it is hoped, the title of this book can be justified in more detail than hitherto). There is a need for effective information systems as the core of any analytical capability. This would commonly be accepted. But we have now added two features: the need to work at a fine *geographical* scale in very many situations, and the need to add *intelligence* by providing the 'what if?' predictive capability through use of the models. Hence intelligent geographical information systems or IGIS.

2.4 OPTIONS

We have already begun to indicate in the previous section how a strategist can begin to use an IGIS to explore options. Here, we offer a preliminary review of some of the further considerations which can be brought to bear. The critical issue to be tackled is the combinatorial problem of design. A design (or an option or scenario, in this case) will be made up of a large number of elements, and there are millions of possible combinations of these. The designer has to have a means of selecting effective combinations. In many situations, a good designer does this apparently by intuition, but probably relying on experience: many years of trial and error have given that person a set of design rules. The same issue arises in the case of the retail strategist who has to 'design' an optimum branch network. The task now is to show how an IGIS can help to provide the 'rules' which will lead to good outcomes.

The SWOT analysis will offer the obvious pointers: defending and enhancing strengths; repairing weaknesses; seizing opportunities; and resisting or responding to threats. The geographical depth of the analysis will allow these questions to be put quite precisely: if it is possible to achieve a market penetration of 20 per cent in a number of small areas, what has to be done to achieve this in the areas where it is only 5 per cent? Experimentation with the IGIS, with new store openings, for example, will provide the answers to such questions.

As experimentation proceeds, it may be possible to develop some principles (or 'rules') which guide the process of finding and testing options. For example, a UK national IGIS contains about 2000 shopping centres, and it will be possible to analyse performance by size of centre and then to find a rule which eliminates all centres below a certain size – though there are always some dangers in this kind of process. It may be possible to distinguish performance and opportunity systematically in geographically different types of centre: town-centre, suburban, out-of-town, or motorway, for instance.

Once the structure of the branch network is determined, more detailed planning issues can be explored using the IGIS; for example: store size and product mix in different locations and situations. By a process of trial and error, a composite plan can be built up. In effect, this process combines the experience and intelligence of the strategist with the analytical and planning capacity provided by the IGIS. It will take only a few minutes to set up and explore an option, so even the initial stages can be accomplished relatively quickly; and then the system can be monitored and updated.

An interesting further question relates to the extent that this human–machine process could be further automated, and software added to the IGIS to increase the level of 'intelligence' by making it search for optimal solutions. It is possible to do this in certain well-defined circumstances, but in most cases, the level of complexity is so high that it is better to take advantage of human intelligence and experience as well. Compromises are also possible, such as asking the computer to assume a green-field situation for the organization as a means of generating a picture of how an ideal 'solution' would compare with the existing one.

The example of retailing has been pursued in this section in order to develop the broad concepts. It is easy to see how similar ideas can be developed in relation to the other examples mentioned earlier. In some cases, the analogy is fairly direct – for example, with hospitals and patients; in others, a little more imagination may be needed to see precisely how an IGIS would be useful. The relevant ideas are pursued in subsequent chapters.

2.5 MANAGEMENT AND PLANNING CAPABILITY

Most organizations have some kind of senior board concerned with the main functions of management: finance, marketing, personnel, estates, purchasing, distribution, and planning, say. There may be separate concerns with information technology (IT) and management information. The chairman and managing director will have the responsibility of co-ordinating the strategic thinking of the board. Problems are likely to arise in this because of the professionalization of functions, which leads to myopia and impermeable boundaries. Specialist information collected in relation to each function is not added to a common pool, for instance. The responsibility for making key decisions may not be clear. Who, for example, is responsible for decisions on the opening of a new branch – estates, marketing, or planning? The implication is clear: information needs to be combined and key decisions which involve several functions should be made corporately. A good management information system is at the heart of the issue in achieving this. The argument of this chapter adds a further element: that in cases such as that of the retailer (and the other examples presented in this book) an IGIS, fully integrated with the main management information system, is a key element.

2.6 THE CONSTRUCTION AND IMPLEMENTATION OF PLANS

If an IGIS is developed for an organization, it will produce the following:

- integrated information availability;
- a battery of performance indicators;
- a predictive capability;
- intelligent, powerful, user-friendly software.

One way of summarizing how this power can be applied is to note its implications for the likely headings in any plan which is produced as a result of the process which has been described. Part 1 of such a plan might consist of 'Objectives, analyses, and options', with something like the following chapter headings:

- objectives;
- statement of performance indicators;
- SWOT analysis;
- analysis of organizational and management structure in relation to locational decision making;
- options to be considered.

There would be some iteration between stages in the work which lead to the production of the plan – for instance, the specification of the performance indicators and the SWOT analysis would lead to the refinement of the statement of objectives.

It should also be recalled that planning is not a neat activity, and this is one reason for separating Part 1 (which can be thought of as a framework for planning) and the associated decision making from Parts 2 and 3, which would be concerned with 'Proposals' and 'Implementation' respectively. While it would be necessary to update Part 1 on a regular basis – certainly annually, and perhaps even six-monthly – Parts 2 and 3 would be changing on a continual basis – for example, for a retailer, as opportunities emerged to acquire particular sites or as the behaviour of competitors changed.

2.7 CONCLUDING COMMENTS

Good corporate planning is now generally recognized as a key feature of organizational success and development. What has been argued in this chapter is that in order for the corporate planner to be effective, he or she needs good information for planning and a position within the organization which enables decision making to be integrated. The IGIS is a critical element in providing the information base, and will also be valuable as an integrator. This is because, although what underpins it in the theory of computer modelling is complex (in order to represent the complexities of the real world), its results can be presented with good graphics in transparent and straightforward ways, which should hold the attention of all the members of the board in presentations and discussions. A large-screen IGIS in the boardroom will be an attractive asset.

It is useful to conclude with a review of the progress which has been made in the development of IGIS in the various sectors which form the subject matter of the rest of the book, and which were listed in Section 2.1. The nature of the progress made depends in part on the culture and politics of particular sectors, and in part on some of the issues of management structure mentioned earlier, especially the integration of information and associated skills.

It is interesting to note that the modelling skills which underpin IGIS were developed 25–30 years ago, while their application in IGIS has been relatively recent, predominantly in the last 10. This illustrates the difficulty of transferring technology – in this case 'soft' technology – from academic research to industry and government. Here the problem has been partly solved by the researchers involved setting up a university company (GMAP Ltd) with this objective. The other side of the equation which determines the pace of advance is the receptiveness of the potential user to the new technology. Since modelling research started in the late 1950s in the context of transport planning and then, more broadly, urban planning, one might have expected local government to be an early convert – and indeed it was, usually with hefty central government support. These studies have become relatively routine in North America, but less so in Europe or the rest of the world. In the UK, the failure to consolidate early gains probably stems from a combination of three factors: first, the early studies were often clumsy prototypes and not entirely successful; secondly, they were being bolted on to a non-numerate town planning culture, and the 'join' was insecure; and thirdly, all forms of planning associated with government became unpopular from about the mid-1970s to the early 1990s, though there are now signs that the tide is turning again (see also the discussion in Chapter 1).

What is perhaps more surprising is that the obvious opportunities for this kind of model-based planning were not picked up by specific departments of local government, particularly in fields like education, where it has been necessary in most areas to re-organize on an almost continuous basis because of demographic and migration factors. Another area where opportunities have not been developed to the full has been in health-service planning, which is the subject of Chapter 6. In the UK, there has been considerable funding of the underpinning research by research councils, central government, and charitable foundations, but it has proved difficult to bolt this on to the agencies of government – perhaps again because they are not culturally disposed to this kind of methodology, even though the efficiency gains from using it are potentially high.

It is perhaps less surprising that, once the skills of organizations like GMAP became visible, the methodology was seized on with enthusiasm by many private sector, market-orientated firms, since there are substantial and measurable increases in profitability from good IGIS-based planning. The difficulties of transfer into these various sectors are probably of two kinds: first, a failure to recognize the potential advantages at all (for what is after all a relatively expensive investment, even if the rewards are high);

and secondly, the fragmentation of management functions in many organizations. IGIS is, as noted earlier, an integrator, and there may well need to be a team approach to corporate planning visible at board level – or at least an effective champion there who can convince others – if the methods are to be adopted. Indeed, a common inhibitor which manifests itself for these kinds of reason is the argument that an IGIS system is impossible because of lack of data. This has never been a problem in practice: it can be solved by integrating the data sources which are usually available in various parts of the organization with those publicly (or commercially) available from external sources. The modelling framework can be used both to achieve the integration and to fill in any gaps.

These gains in the market sector have been predominantly in areas which are, in a general sense, concerned with retailing, including the motor and financial service industries as well as more conventionally defined, shopping-centre-based retailers. There is considerable scope for similar applications in other industries, the core of the activity being the integration of data on flows: from sources to production processes and from production to customers. One project has been carried out, for example, by GMAP in the chemical industry.

Even in organizations where IGIS has been successfully developed, the time horizon for planning has typically been very short-run – usually the immediate future, a year or two ahead. It would be potentially valuable to integrate IGIS with the other aspects of forecasting mentioned briefly in Section 2.3: longer-run demographic, economic, and geographical scenarios. For all the uncertainties of such exercises, and recognizing that they would need continuous re-appraisal as events unfolded, they might none the less add a significant new dimension to current short-run-orientated practice.

The future development of IGIS is interestingly poised. There have been sufficient studies carried out to demonstrate feasibility and effectiveness, and yet the range of application, both within sectors where the best examples are currently located and in those where there has been barely any development, could be much wider.

CHAPTER 3

GIS and spatial decision support systems

3.1 INTRODUCTION

When the history of academic geography in the 1990s is written, the development of GIS may be seen as the single most important occurrence in the discipline during the decade. It has moved geography to centre stage as an applied, technologically based discipline. It has generated a plethora of new employment opportunities, books, journals, courses, and even companies. Yet, to date, GIS has received scant attention in texts which are supposed to offer students a guide to the major events and paradigms of the discipline (Johnston 1991, Cloke et al 1991: though this is at last beginning to change – see Taylor and Johnston 1995). Perhaps this is in part a reflection of the fact that the majority of significant developments in GIS have been undertaken outside the geography discipline by computer scientists, computer cartographers, and so on (see Openshaw 1989a). This has slowly changed as research funds have been pumped in to the National Centre for Geographical Information (NCGIA) in America and the Regional Research Laboratories (RRLs) in the UK, yet the technical contribution of practising geographers is still limited. What has remained uniquely geographical is the problems that GIS have been developed to address. This problem-solving focus and related approaches are recurrent themes throughout this book.

That said, the aim of this chapter is to explore the potential for GIS in terms of its ability to contribute to the solution of business problems in both the public and private sectors, building on the framework described in Chapter 2. In particular, we are concerned with those business problems that have a distinct geographical component. The main argument we shall develop is that, although GIS provides an important and useful first stage of data handling and presentation within the problem-solving process, there is a need to improve considerably the level of spatial analysis contained in the process if businesses are to embed a GIS capability at a strategic level. Section 3.2 looks at how GIS is currently defined and at the types of application area where it has to date been successful. Section 3.3 deals with what a GIS can achieve in terms of spatial analysis and what the blueprints for the future look like. Section 3.4 offers a critique of proprietary GIS on the grounds that many of the real-world spatial problems are more complicated and require further analytical skills than is allowed by traditional GIS. This is not to deny the value of current systems, but simply to argue that the reservoir of spatial modelling methods developed over the last 30 years needs to be tapped in order to make the power of GIS match the problems they are being pitched against. This leads us on to examine the concept of spatial decision support systems (SDSS) or IGIS in Section 3.4 as an alternative if complementary approach. We suggest that these SDSS should be based on modelling skills built up over many years, which offer a unique range of support power to the modern public or private sector planner. The chapter concludes with an illustration of both the power and shortfalls of modern GIS in relation to SDSS through the example of the use of models in facility location planning.

3.2 GIS: DEFINITIONS AND FUNCTIONALITY

There are now a large number of good introductory texts on GIS presenting a wealth of alternative definitions (Maguire et al 1991, Martin 1991, Huxhold 1991, Worrall 1990a, Scholten and Stillwell 1990). To appreciate the aims of GIS it is useful to dissect the main components of its title. 'Geographical' relates to the fact that we are interested in data and attributes which have some sort of spatial identity. These might be point features relating to specific sites such as schools, hospitals, or shopping centres; line features such as rivers or roads; or area features such as Census wards, parliamentary constituencies, or postal districts/sectors. 'Information' is a more difficult concept to summarize. It is usually argued that what defines something as information is its usefulness in decision making or planning. As we shall see later, this may be present through existing data sets or created through data linkages or data models. Thus as Lucas (1978:5) writes: 'An information system is a set of organised procedures which, when executed, provides information to support decision-making.' *Geographical* information systems, therefore, should possess these properties in relation to geographical data and information.

The history of GIS development is well told by Coppock and Rhind (1991). They describe the development of GIS from the first application in Canada in the 1960s to the present day, highlighting significant developments such as the arrival of super-mini computers and the incorporation of relational database management system technology in the 1980s. Much of the technical advance has been made in the United States, especially with regard to the development of proprietary systems. In the UK, a major political landmark was the government enquiry into geographical handling in the mid-1980s and the subsequent publication of the Chorley Report (Department of the Environment 1987). This Report recommended greater investment and use of digital maps, more investment in academic research and training (leading to the setting up of the Regional Research Laboratory Initiative – see Masser and Blakemore 1991), and the encouragement of new organizations which would bring together interested parties from all walks of life (leading to the setting up of the Association of Geographic Information).

Today the GIS industry is a multi-million-pound business serving a vast range of diverse organizations associated with spatial data handling, and is threatening to become a major branch of academic geography in its own right. Examples include all aspects of human and physical geography, and the text of Maguire et al (1991) is testimony to this effect. The potential of GIS is best illustrated through an examination of what it can offer the geographer or planner, whether public or private sector, in terms of spatial analytical functions. We will review these functions under five major headings.

3.2.1 Data storage, retrieval, and display

Martin (1991) argues that GIS has largely developed from computer-assisted cartography and the remote sensing/environmental science worlds, and it is perhaps not surprising that issues of data storage, retrieval, and display are still at the forefront of many GIS applications. Indeed, Rhind et al (1991:314) recognize that 'GIS seem too often limited to mapping, information management and simple inventory', and Landis (1993) estimates that for 90 per cent of present users the most important GIS capability is presentation mapping. That said, it is apparent that there are some very impressive examples in the literature spanning a variety of different systems of interest. Clearly, one of the major advantages of GIS for many beginners has been their ability to store and integrate different data sets within a single system. There is often nothing particularly innovative about this, but the perceived benefits are significant.

On the environment side, particularly impressive examples of GIS use primarily for data storage, retrieval, and display include the United States Geological Survey (Starr and Anderson 1991), the CORINE environmental database of the European community (Mounsey 1991), and the various soil information systems described in Burrough (1991). Within human geography, similar examples can be found in urban information systems covering land use, population groups, employment characteristics, and housing details (Parrott and Stutz 1991, Huxhold 1991, Dale 1991). Similarly, many of the utilities have powerful digital mapping and storage systems to replace wholly inadequate paper records (Mahoney 1991, Rector 1993). A number of these important information systems have been placed in the public domain to ease the access to a wide range of useful data sources. These include projects such as Domesday (see Openshaw et al 1986), which includes most official statistics concerning the UK (among them the 1981 Census), a national land-use survey, and over 22 000 maps and 40 000 photographs of life in Britain in the mid-1980s. A second well-known example is National On-Line

Manpower Information System (NOMIS) (Blakemore 1991), financed by the Department of Employment and including information on populations and households (through the Census of Population), jobs (through the Census of Employment), and vacancies and unemployment (through JUVOS data, NOMIS unemployment and vacancy date). This is available for the whole of Britain on a wide range of spatial scales.

All these systems share the characteristic of holding enormous data banks on a large number of related components with the facility to retrieve and map/graph any of the information the user desires. This facility is particularly important, as the map enables the user to 'humanize' the results of spatial queries and aids the process of pattern recognition. The expanding field of computer cartography has allowed GIS users to be more ambitious in relation to visualization (see Buttenfield and Mackaness 1991). These improvements have come in many forms, from computer movies to sophisticated digital terrain representations particularly useful for the reconstruction of any landscape for which data are available (Howes and Gatrell 1993, Weibel and Heller 1991). These terrain models have become vital training systems for military use. Intergraph, one of the largest players in the digital map industry, are responsible for the guidance systems in cruise missiles, whilst both McManners (1992) and Smith (1992) describe the importance of GIS and terrain modelling in planning of military strikes during the Gulf War of 1991. (We shall return to the ethics of this in Section 3.3.)

3.2.2 Data linkage

Data linkage is one of the most fundamental methods of adding value to data. In most GIS the concept of *polygon overlay* is central to the process of linkage. Figure 3.1 gives a graphic illustration of the overlay procedure. Data are stored in layers which can be retrieved from a GIS and then overlaid one on another to answer questions such as which sites are most suitable for development of various types. In Figure 3.1 we can see how a site for development is determined by checking the suitability of the location on each of a number of key criteria. The locations which pass the test on all of these criteria are then suitable for development. Hence the process is one of elimination by seeking to reduce the 'possibility set' for some kind of development. Dangermond (1983) uses the overlay procedure to select new town sites in southern California by combining large regional databases on a variety of physical and cultural attributes. Similarly, Cowen et al (1983) provide a 27-layer data file for the purpose of minimizing land-use conflicts of new route development in Georgetown, South Carolina, USA. Priorities may be attached to different criteria by using *weighted polygon overlay*. This simply allows one criterion to have more importance than others in this elimination process.

The overlay procedure is often undertaken in conjunction with *spatial buffering*, another standard feature of all proprietary GIS. Buffering enables

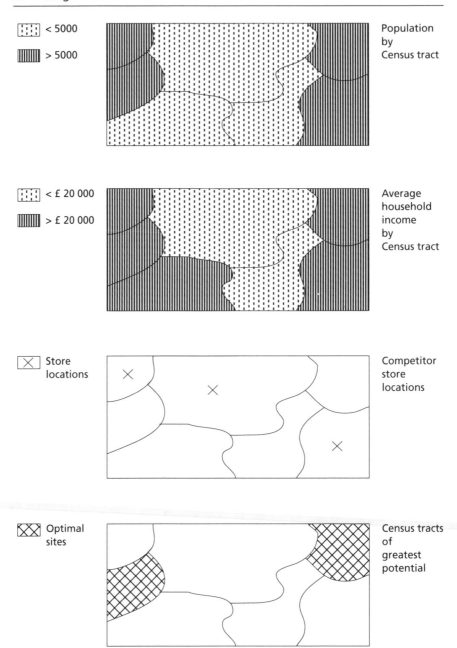

Figure 3.1 The overlay procedure in GIS used to find optimal locations for new retail business

the user to determine an area at a chosen distance from either a point location (school, hospital, nuclear power site, etc.) or a line feature (river, road, etc. – see Figure 3.2). Typical queries might include the calculation of population at risk from a tanker spillage within two miles of a major motorway or trunk road, or how many compensation claims might result from aircraft noise pollution in a fixed spatial corridor surrounding an airport flight path. We shall look at further illustrations of these in relation to facility location in Section 3.2.5.

3.2.3 Geocoding and geosorting

Once data has been geocoded (that is, given a spatial reference) then the GIS can perform a number of spatial queries. This is also known as geosorting. For example, once a specific map location (such as a school, motorway, or parcel of land) has been highlighted, the GIS will search through all its information banks to retrieve all recorded information associated with that geographical feature. Conversely, the entire databank can be searched to find all attributes associated with a certain spatial reference point. Landis (1993) notes that geocoding can be effectively linked to overlay and buffering procedures to answer more sophisticated queries. One such approach is labelled *point-in polygoning*:

> Suppose a furniture store chain wants to inform its regular customers about the location of a new, nearby store. By geocoding its customer addresses to a street network and then geosorting these points within a 5-mile radius of a new store, the company can more effectively reach its customer base and limit the size of its mailing.
>
> (Landis 1993: 34)

This process is illustrated in Figure 3.3. which shows average house prices in Berkeley, CA, by census tract. The individual houses (points on the map) are located within the Census tracts (polygons) to produce the overlaid thematic map.

3.2.4 Network analysis

This is one of the most frequently used components of a GIS and underpins much of GIS use in the utilities and transportation. Once a road system, a river catchment basin, or a pipeline system has been digitized, a GIS should quickly be able to answer questions such as 'What is the quickest way from A to B?' (routing). The response algorithm can be made more complicated by adding *impedance* on to major routeways. These may include speed restrictions, one-way systems, or temporary blockages caused by accidents or road repairs. Another important use of network analysis is the allocation of resources to predefined centres, normally based on minimizing total distance travelled. In a sense, this operates as a form of simple catchment-area analysis. For example, the number of pupils in a district can be assigned to the nearest network link and then in turn allocated to schools (centres) based on user-defined rules concerning maximum travel times. Assigning a centre capacity provides further

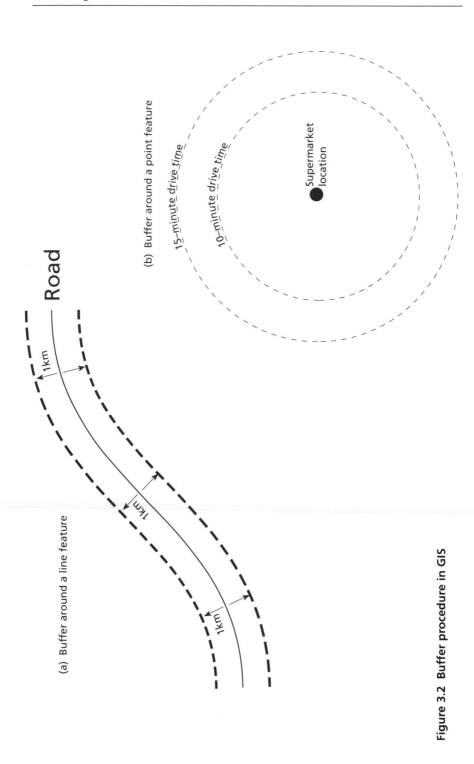

(a) Buffer around a line feature

Road

1km

1km

1km

(b) Buffer around a point feature

15-minute drive time

10-minute drive time

Supermarket
location

Figure 3.2 Buffer procedure in GIS

realism to be built into the allocation procedure. Figure 3.4 summarizes the simple network operations of routing and resource allocation.

Badillo (1993) provides a comprehensive list of application areas of network analysis, many of which are making headlines in the media as major contributions to more efficient traffic planning. These include car navigation systems designed to give up-to-date information as to how to avoid the latest roadworks or accident hold-ups, and *global positioning systems*, which keep precise tabs on the locations of fire engines or ambulances so that the nearest vehicle can always be dispatched and response times improved (see Ward 1994). Coupled with navigation systems, these allow the drivers to find the quickest and safest routes to accidents or emergencies given the latest local traffic conditions. Peel (1993) also provides a useful case study of the use of a GIS to ease traffic congestion in the heart of London by prioritizing flows during peak periods.

When these are combined with other GIS features, such as buffering and overlay, it is possible to produce sophisticated 'models'; for example, to

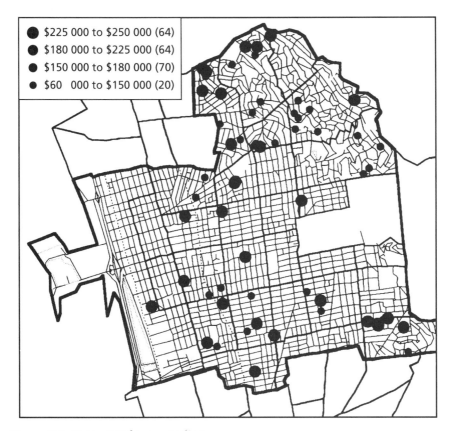

Figure 3.3 Using GIS for geocoding
(*Source*: Landis 1993: 36)

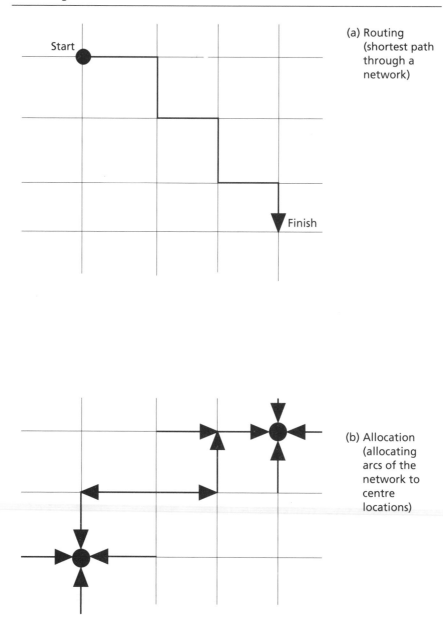

(a) Routing (shortest path through a network)

(b) Allocation (allocating arcs of the network to centre locations)

Figure 3.4 Routing and allocation using network analysis

direct evacuation procedures in emergencies. Gatrell and Vincent (1990) describe their work, undertaken with Cumbria police, to look at the population at risk in different localities at different times of the day in relation to hazardous industries (using buffering and overlay), and to combine this with a detailed road network of Cumbria which would facilitate the planning of fast evacuations (network analysis). The fact that a seminar at Lancaster devoted to emergency planning in 1991 attracted more than 300 delegates is testimony to the importance of such applied GIS.

3.2.5 Spatial analysis and GIS: the case of facility location planning

The level of spatial analytical techniques available in GIS is one which has increasingly concerned academics, and we shall return to this issue in

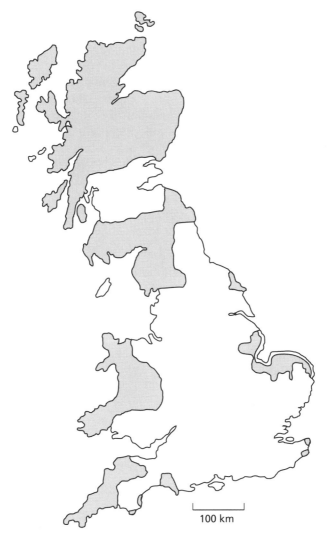

Figure 3.5 Optimal locations for nuclear power stations (best 100 000 1 km grid squares)
(*Source*: Openshaw 1986: 303)

Section 3.3. However, it is useful first to see how the functions described above can be and have been combined. Since the theme of this book is facility location in its broadest sense, we shall focus on how GIS has been used to aid decision making in three related areas: hazardous waste disposal, the location of fire stations, and retail site selection.

3.2.5.1 Hazardous waste sites or nuclear power sites

In many respects this area of policy analysis is one of the most impressive applications of the combined use of data storage, retrieval, graphics, buffering, overlay, and network analysis. The aims generally have been to find sites which maximize or minimize certain features affecting safety, in relation to the siting of noxious industries or waste disposal sites. Openshaw (1986a) provides a good example of buffering and overlay in searching for the safest locations for nuclear sites, primarily on the basis of the remoteness of coastal locations from areas of population. The results of this exercise are shown in Figure 3.5. Similarly, Siderelis (1991) lists a number of 'rules' required for optimizing the location of hazardous waste disposal sites in South Carolina. These are shown in Table 3.1. Scanning this list, it is possible to see how GIS could use buffering and overlay to eliminate land areas which violated this rule base (see also Gatrell and Vincent 1990). Although this list seems long and complex, GIS can find any occurrences of the violation of these rules with immense speed. It is likely that the data entry and creation of the topology would take the bulk of time in constructing the GIS.

Table 3.1 Abridged rule sets for finding the optimal location for a new hazardous waste site (*Source*: adapted from Siderelis 1991)

Development must not take place:

– upon an inland lake;
– upon an upland bog or pocosin, or upon marsh or swamp;
– in a coastal hurricane storm surge or inundation area;
– within 40 km of an existing hazardous waste site;
– more than 60 km from the interstate highway;
– within 0.4 km of the epicentre of a seismic event > 3;
– in or on an Indian reservation or federal military reservation;
– in or on state and national parks and forests;
– within 8 km of the state boundary;
– upon certain geological formations;
– upon general soils liable to flood;
– within the corporate limits of a municipality;
– upon a surface slope greater than 15%;
– upon protected fresh-water or tidal salt-water areas;
– within 3 km of surface water supply used for drinking.

3.2.5.2 Location of fire stations

Although the traditional domain of location-allocation models (see Section 4.3.6.2), the siting of fire or ambulance stations has been another classic use of GIS for facility location. Parrott and Stutz (1991) illustrate this for San Diego. The first step is to buffer a series of travel-time isochrones around potential fire stations within the city. Within each buffer the population at risk of fire (household and workplace) can then be calculated by the overlay routine. The Allocate routine, in this case in the proprietary GIS package ARC/INFO, can then be used to assign each street segment to the nearest fire station on the basis of travel times. It is then reasonably quick to test a variety of scenarios regarding fire-station locations to minimize the overall response time to serve the city's population. Table 3.2 shows a sequence of two-minute response-time bands for each of a number of alternative fire-station locations within La Mesa in San Diego County. Within each time band are listed the 1991 population, housing units, and population in employment. Clearly, the sites which can reach the population at risk in the shortest response times are the most favoured.

3.2.5.3 Retail site selection

A third application area has been in the retail and marketing environment. Again, we shall simply describe the traditional approach here, but we will

Table 3.2 Changes in catchment-area populations for different fire-station locations (*Source*: adapted from Parrot and Stutz 1991)

	Number of stations	Time from station(s) (minutes)	Population served	Remarks
Model A	1	<2	14 712	Worst-case scenario
		2–4	37 645	
		4–6	1 436	
		Total	*53 793*	
Model B	2	<2	19 904	Improvement on A
		2–4	32 918	as the number of
		4–6	972	stations grow
		Total	*53 794*	
Model C	3	<2	23 132	Existing condition
		2–4	29 749	and best-case
		4–6	913	scenario
		Total	*53 794*	
Model D	3	<2	20 981	Relocation of sites
		2–4	31 633	but no improvement
		4–6	1 180	on C
		Total	*53 794*	

return to this example later for a more detailed critique of the use of GIS for this sort of facility-location exercise, and in later chapters we will show how intelligent GIS-based procedures are much more powerful.

The standard procedure in retail analysis would be to set up the required database of populations within cities or regions. A new site would be selected with the aim of using the GIS to calculate likely revenues for the new store. Once the information is stored in the GIS, the user can buffer travel times around the new store and then calculate the population within each time band using the standard overlay procedure. This is illustrated well by Beaumont (1991a, b), Howe (1991), Elliott (1991), and Ireland (1994), and shown graphically in Figure 3.6. Once an estimate has been made concerning the demand within the likely catchment areas (normally used in conjunction with a market survey to see how far people typically travel to a similar store elsewhere in the corporate chain), then a variety of methods may be used to translate population totals into branch sales. The most likely method is *fair share* (Beaumont 1991b), because of the great difficulty of dealing with the competition using these sorts of method. Hence, if there are three other competing stores in the buffered catchment area of the new

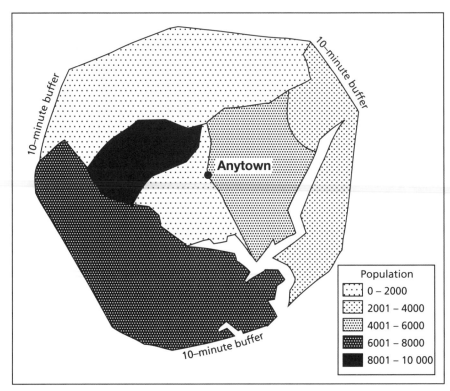

Figure 3.6 Combining overlay and buffering to calculate catchment-area population for a new store in 'Anytown'
(*Source*: adapted from Elliott 1991)

store, then the new store may be expected to obtain 25 per cent of the revenue generated in that catchment area. This simple fair-share allocation could be weighted by store size or by retail brand to increase realism. The alternative is to assume the consumer will travel to the nearest store within the catchment area (*dominant store analysis* – see Ireland 1994).

Elliott (1991) illustrates this well. She notes how a GIS can be used to calculate the population within a 20-minute drive time for any new or existing Debenhams store. The great difficulty with this procedure, however, is that it does not allow for the complex set of real interactions between residential areas and retail locations which are distorted by intervening opportunities. As Elliott herself acknowledges, the presence of competing centres will restrict the catchment boundary of a new store in certain directions. Her response is to 'override the drive time where it seems appropriate' (1991: 171). Such subjectivity is the precise reason why methods like this are not as accurate as alternative ones (see Chapter 4).

3.2.5.4 Concluding comments

These three application areas illustrate the typical use of GIS for facility location and the type of spatial analysis currently available to address such issues. Although the details above have been kept brief, the reader will be aware of the potential applied importance of GIS.

It is clear that GIS (and geodemographics – see Chapter 4) has whetted the appetite of many public and private sector planners in relation to the importance of spatial analysis. At the same time, historians of quantitative geography will be well aware of the struggle of that sub-discipline to remain applied and relevant throughout the 1970s and 1980s, as more and more planners became disillusioned with models and their outputs. It was a time when geographers by and large turned to alternative methodologies, normally based on 'radical' perspectives, and left quantitative geography increasingly in the hands of regional scientists, with their all-too-frequent preoccupation with abstract isotropic plains, two- or three-household case studies, and an often massively complex set of equations (see the critiques of Beaumont 1986, Openshaw 1986b). Although the mainstream geographical community remains 'qualitative', the increasing appearance of GIS has rekindled interest in applied spatial analysis across a broad range of institutions outside academia. This is partly reflected in the range of applications described above, and also in the membership lists of organizations such as the Association of Geographic Information. In turn, these developments have helped to rekindle interest in other forms of spatial analysis, such as modelling, by allowing modellers to connect to the main advantages of GIS – data storage and graphical output – and present the outputs of their models in a way that users can understand and interpret. (Section 3.3 considers other reasons for this renewal of interest.)

The potential of GIS is clearly large and many have been undoubtedly successful. For every success story, however, there are a number of victims: customers who have had their fingers burnt quite badly through being seduced by the glossy demonstration disks which often make GIS construction look so simple. Many purchasers simply lack an understanding of the large-scale programming effort which is often required even in the most user-friendly of GIS packages.

3.3 GIS AND SPATIAL ANALYSIS

In this section, we offer a sympathetic critique of spatial analysis within GIS and, in particular, within proprietary systems, acknowledging the considerable advantages demonstrated through the types of application described in Section 3.2.5.

The first major concerns come from mainstream geographers, many of whom regard GIS as simply the rebirth of the type of quantitative geography apparently discredited during the 1970s. In addition, Openshaw (1989a) points to the fact that since GIS is an overtly applied discipline it may give the impression of not being particularly scholarly. The debate between Openshaw (1991a, 1992a) and Taylor and Overton (1991) in the editorials of *Environment and Planning A* (1991) is a good illustration of these sorts of argument. What is also apparent is the prejudices of those who continue to label any sort of quantitative analysis as 'positivist'. Lake (1993: 405, 407) provides a good example of this. Not only does he request GIS developers to 'relinquish their positivistic assumptions' and 'respect the subjective differences among the individuals who constitute the irreducible data points at the base of the GIS edifice', he also questions the very ethics of GIS, particularly through the emotive example of its use in the Gulf War of 1991 (see also Bondi and Domash 1992, Pickles 1991, 1995, for general anti-GIS sentiments). It is difficult to counter such totally opposing views. One could get embroiled in the detail of individual arguments (such as whether Lake would prefer a 'conventional' five-year war with all its resultant casualties, or whether Lake has really thought through the relationship between positivism, GIS, and quantitative geography – see Bennett 1985, Wilson 1989). However, it is more constructive to continue to illustrate the importance of these methods in applied human geography. Fashions in academic geography have changed dramatically over the last 30 years and will continue to do so. What remains constant is a need for better methods of problem solving in the broad planning and business communities.

Perhaps of greatest concern to the GIS community is the lack of flexibility within existing packages and the fact that many users find them difficult to operate because of their associated computer languages. Spatial analysis is a messy concept, and it is often difficult at the start of an exercise to work out fully the exact course of action over a project's life span. Indeed, it is often difficult to know what the exact objectives are at the start of such an exercise. This in itself leads to frustration when packaged solutions are flexible enough neither to answer strategic planning questions from the outset nor to change direction over time. This is the result of advances in GIS being technology-led rather than a reaction to an understanding of the ways in which geographical analysis can help solve problems in business and commerce. Perhaps it is also the result of GIS being 'oversold', in that vendors are quick to argue that anything is possible, supported by glossy demonstration disks which can be all too seductive (see also Cross 1991, Openshaw 1993, Aangeenbrug 1991).

The requirement to become involved with computer programming also produces an air of user-unfriendliness. Even though proprietary systems can be 'customized' (using macro languages to build structured menu systems), there is still a great deal of programming skill required to operationalize a proprietary GIS package successfully. For example, it is suggested that it needs about six months' training for an average person to have a full understanding of the command structures in ARC/INFO (although Cresswell 1995 argues that user-friendliness in proprietary GIS generally is improving rapidly).

The major cause for concern with existing GIS, however, is the level of spatial analysis functionality currently on offer. This has been the theme of a large number of GIS seminars and conferences. Openshaw (1991) goes as far as saying that there is no 'real' analysis except in a data description or cartographic sense. From the wealth of published material it seems there are three main ways of trying to improve the level of analysis in modern GIS. The first simply involves a greater number of functions being available through proprietary packages. This is clearly the ESRI route through its leading package ARC/INFO. Maguire (1995) gives a checklist of features added to the package over the years and of plans for future additions. The list is impressive in one sense, but many of these items will be seen as peripheral (such as many of the statistical routines) and in many ways old-fashioned, especially by those in the business and commercial sectors. The impression is that such additions have been largely the result of academic pressure from a body of statistical geographers, since spatial analysis is still regarded by many as equating exclusively with statistical analysis. There has already been some progress in linking GIS to statistical routines (Ding and Fotheringham 1992, Kehris 1990), yet the demand for more remains. Flowerdew (1991a) notes that an important conclusion of a 1991 workshop on GIS and spatial analysis was the urgent request for the calculation of Geary and Moran co-efficients and Kernal estimation techniques to be added to proprietary systems (see also Bailey 1992, Anselin and Getis 1993, Griffiths 1993, Goodchild 1987, Goodchild et al 1992). Adding more features in this way might seem harmless enough (the 'you do not have to use them if you do not wish to' argument), but it is important to reflect that adding extra modules to black boxes such as ARC/INFO might, whilst solving some analysts' plea for inclusiveness, lead increasingly to the use of such methods in an environment of user-ignorance of the meaning and interpretability. This is less of an issue if the work is controlled by persons with good spatial analytical skills, but unfortunately is a definite problem if such procedures are left to novices. In addition, such a strategy continues the technology-led rather than the solution-led approach to spatial analysis.

A newer theme in the GIS literature is how the link between GIS and spatial analysis can be transformed to ask not 'How can additional spatial analysis help improve the performance of GIS?' but 'How can GIS functionality actually improve our knowledge of exploratory data analysis?'. This is the theme adopted by Fotheringham (1994) and Fotheringham and Xie (1995). Prior to this work, many authors envisaged the real advantages of linking GIS and spatial analysis routines as coming from the greater visualization opportunities that the former provided (see Section 3.4). However, Fotheringham points out that GIS technology can in itself help users understand better some classic geographical problems, such as sensitivity of analytical results to zone definition, the nature of spatial non-stationarity, and the definition of spatial outliers. These are two important new papers in the debate on GIS and spatial analysis.

A second route, favoured by Openshaw (1991, 1992), argues for a new era of spatial analysis based on a more flexible approach to finding spatial patterns and clusters. The idea here is to abandon any *a priori* theory or model to account for spatial patterns (indeed, Openshaw advocates that all previous spatial analysis textbooks be incinerated!) and allow new computer software to search for such patterns or clusters without any theoretical preconceptions. Such software contain libraries of 'pattern exemplars' or generic algorithms for pattern analysis. The motivation for this sort of approach lies in the huge volumes of data currently available and the urgent need for new software to explore all possible patterns and correlations: 'Put another way, the historic emphasis on deductive approaches is becoming less practicable because of the increasing dominance of data-led rather than theory driven questions' (Openshaw 1992: 2). There have been some impressive successes using this sort of approach, particularly in the well-documented cases of Openshaw's work on finding cancer clusters (Openshaw et al 1987, Openshaw and Craft 1991).

At this moment it is difficult to see how such generic tools can be made more widely available to the business community, given the hardware requirements of the software and the often specific location problems many businesses actually face. A more important issue may be whether such models have any explanatory power. By creating new models which fit the data specific to one period, how confident can we be that the model is robust enough to replicate or forecast changes when the data are altered in some small way? At least some of the older models do have a history of successful application, as we aim to demonstrate in this volume. Thus, we would agree that Openshaw's methodologies are a valuable addition to our armoury, but not at the expense of committing all existing methods to the museum case.

A third route is to move towards integrating GIS with other software through direct or indirect coupling. This involves providing specialist software outside the main package but linking in the data storage and graphics of the GIS through new command structures. This approach forms the heart of *decision support systems* or, when applied to geographical data and problems, *spatial decision support systems* (SDSS). The definitions and history of their development from management schools are supplied by Densham (1991) and Densham and Rushton (1988). They reiterate the fact that spatial problems are generically complex and are usually ill defined or semi-structured. For these reasons we need systems which can handle a large variety of different business problems in a user-friendly environment (that is, with no visible computer commands other than a start procedure). Typically, GIS is not enough here because of the lack of such suitable analytical functions. What is required is the ability to link spatial models with the strengths of GIS in storage, retrieval, and graphics.

This is a message increasingly echoed elsewhere. Batty (1993) and Batty and Xie (1994a, b) outline the benefits of running and calibrating urban models within a GIS framework, emphasizing the importance of visualization in aiding the interpretation of both calibration statistics and model results.

The rest of this chapter outlines the importance of adding spatial modelling and the tasks it allows us to tackle. The remainder of the book, in effect, provides working examples.

3.4 LINKING MODELS WITH GIS

We have attempted to review a broad range of application areas in the socio-economic arena and have suggested that, as currently perceived, GIS may fulfil only a limited number of potential user requirements. In this section, we outline how a marriage between model-based methods and techniques from GIS can result in a powerful analytical system that meets these requirements. Most of our illustrations are kept for subsequent chapters: here we discuss what types of role models can perform and how these roles relate to tasks that would enhance a typical GIS.

A major research question at the outset is how to link GIS with models in the most effective manner. One proprietary GIS (ARC/INFO) has already begun this process by incorporating spatial interaction and location-allocation models into its network analysis module. This direct coupling is to be broadly welcomed, but it does raise concerns regarding the form of the models available to the user (which are limited to the most basic of model formulations) and the safety of calibration for those lacking training in modelling methods. Research on the robustness of these new routines remains an ongoing task (but see Benoit and Clarke 1995).

An alternative route is to offer pathways through current GIS menu systems to models which have been developed and calibrated off-line. The user thus has access to more powerful analytical routines but from the safety of the user-friendly, menu-driven package. Data transfer to and from the modelling environment is thus seamless, and model outputs can be transferred back into the GIS for mapping, aggregation, overlay, buffering, network analysis, etc. Such indirect coupling is a key feature of the difference between standard GIS and IGIS or SDSS.

We shall now focus on what models can offer beyond the capabilities of existing GIS.

3.4.1 Transformation of data

A basic role of both GIS and models is to provide a framework within which data can be manipulated and transformed. In GIS, this can take the form of most simple mathematical operations, which will take different items of data and produce new information. This is best illustrated by an example. Let us assume that a motor manufacturer has details of sales of new vehicles by postal district and the total sales of all vehicles in each district. These two items of data can be transformed into an interesting item of information, namely market penetration by postal district, by dividing manufacturer's sales by total sales. This will be widely illustrated in Chapters 5 and 9. However, if such market-share data is not available from published sources then models are required to estimate it. This involves estimating likely flows from origin zones to service outlets through the use of spatial interaction models (see Chapter 5).

A more detailed approach to data transformation arises through the comprehensive use of model outputs which have been articulated as a system of spatial performance indicators (Bertuglia et al 1994a). For many services provided, both publicly and privately, two fundamentally different types of performance indicator can be developed. The first can be termed 'facility-based', and relates to the efficiency and effectiveness of the outlet in relation to its catchment population. Catchment populations can be considered as the notional number of individuals that a facility is serving irrespective of where these people live. It can be calculated in a number of different ways from spatial interaction data. Once calculated, it can be used as the denominator in calculating performance indicators such as the facility expenditure per head of catchment population. The second type of performance indicator focuses on residential districts. Here a typical calculation may involve the notional number of hospital beds supplied to the residents of district i.

Through the use of a battery of performance indicators, a good picture of service supply and utilization can be built up. In addition, transforming model outputs in this way allows the results to be more attuned to key planning problems of the day (Bertuglia et al 1994a). We shall explore such performance indicators for the private sector in Chapters 4 and 5 and for the public sector in Chapter 7.

In our experience, this type of model-based data transformation and spatial representation of performance indicators is of enormous value to many managers and planners: it lets them see data and information in a way they are not accustomed to, and to gain new insights. Importantly, it provides a starting point for the application of more sophisticated methods in later stages of work.

3.4.2 Synthesis and integration of data

One of the main points emerging from the Chorley Report (Department of the Environment 1987) is the extent to which data are collected and assembled at different levels of spatial resolution using different categories to classify variables, such as age, social status, and so on. As we saw in Section 3.2, GIS is a powerful instrument for integrating these different data sets through overlay techniques. An important contribution to the overlay procedure would be to add the results of the sort of analysis described in the previous section. That is, a performance indicator framework would not only provide new ways of transforming data but would itself create a large pool of information that the analyst might find difficult to comprehend fully. An IGIS solution to this problem might be to overlay model-based performance indicators in much the same way as conventional GIS data overlay procedures.

However, for many variables, such as income, information is not collected at anything like the required level of spatial disaggregation to make it useful. In these cases, models can be employed to useful effect – linking and merging data files and making best estimates of missing information. Traditionally, methods such as entropy maximizing have been used to estimate variables at the aggregate level (cf. Wilson 1970). A new way of achieving the same effect, but typically incorporating much more detail and a larger number of variables, is through *microsimulation*. Here households and constituent individuals are recreated from different types of data set and specified at the highest available level of spatial resolution – the household itself or small groups of households in the form of enumeration districts. These households and individuals can then be readily aggregated to provide cross-tabulations of any relevant variables at any upwardly compatible spatial scale. The methodology is underpinned by a procedure known as iterative proportional fitting (Fienberg 1970). A couple of illustrations will help to clarify the procedure.

Whenever published tables are produced from the Census of Population, there is clearly a limit to the number of variables that can be shown, and hence the interdependencies between variables is correspondingly limited. Though some tables may have a three- or four-fold classification, the majority of tables are two-dimensional. For example, it is possible to obtain

Table 3.3 Household and individual attributes created in Leeds micro-database (*Source*: adapted from Birkin and Clarke 1988)

Attribute	Number in category	Details
(a) Household attributes		
Location (enumeration districts)	1565	1. DAAA01
		2. DAAA02
		:
		1565. DABK47
Household structure and composition	5	1. Single person, retired
		2. Single person, not retired
		3. Married couple, no kids
		4. Lone parent family
		5. Married couple with kids
Household tenure	3	1. Owner–occupied
		2. Council rented
		3. Other
Country of birth	7	1. Great Britain
		2. Eire
		:
		7. Rest of the world
(b) Individual attributes		
Status within household	5	1. Head
		2. Spouse of head
		3. Child of head
		4. Other, dependent
		5. Other, non-dependent
Exact age		0–85+
Sex	2	
Marital status	3	1. Married
		2. Single
		3. Widowed or divorced
Economic activity	3	1. Inactive
		2. In work
		3. Retired
Socio-economic group	7	1. Employers and managers
		:
		7. Others
Industry	7	1. Agriculture
		:
		7. Other services

unemployment rates by any of the following variables: sex, age, ethnic status, and occupation. However, if we wished to calculate the number of 16–19-year-old, non-British, unskilled male workers currently unemployed we would usually have to undertake some kind of survey. The alternative is to generate a joint probability distribution for this attribute vector and then synthetically create or extract individuals from this (see Birkin et al 1990). The modelling skill is to build up one attribute at a time so that the probability of certain attributes is conditionally dependent on existing ones.

Good illustrations of this work are provided by Birkin and Clarke (1988) and Birkin et al (1994). They generated a micro-database for the Leeds Metropolitan District based on five household and seven individual attributes. These are shown in Table 3.3. The power of this database is its flexibility in creating arrays based on the combination of any of these disaggregate variables.

In addition to the problem of the limited joint distributions in the published Census there is the problem that some information is not available, for the simple reason that no question was asked on the topic. Perhaps the most important of the missing variables in the UK are household incomes and expenditures (the survey of Marsh et al 1988 identified an urgent need for indicators of income and receipts of benefits). However, these are available for a sample of households through the New Earnings Survey and the Family Expenditure Survey respectively. The problem is simply how to link such survey data effectively with the small-area Census statistics. Clarke (1986) showed that it was possible to construct an income-generation module based on the industrial sector and occupation of the individual from the New Earnings Survey. The next step is to examine the small-area breakdown of occupation groups by industry type and build up probability distributions of household incomes (see Birkin and Clarke 1989). We will describe such work in more detail in Chapters 7 and 8.

3.4.3 Updating information

One of the clear deficiencies of any data set, particularly the Census of Population, is that it rapidly becomes out of date. No GIS has so far been able to offer users an effective on-line method for updating spatial data. How much of a problem this may pose depends upon the particular application. Between 1981 and 1991 increasing unemployment, council house sales, and counter-urbanization quickly changed the nature of the urban environment in many UK cities and regions (Champion et al 1987, Champion and Townsend 1990). Thus, as we approached the late 1980s the reliability of Census statistics became more questionable. The same will happen as we approach the late 1990s in relation to the Census of 1991. However, some information is collected annually on these trends in the UK, mainly indirectly, through electoral registers, the Post Office Address File, Regional

Trends, the Family Expenditure Survey, the New Earnings Survey, and various Office of Population, Censuses and Surveys (OPCS) population monitors. These tend to be available only at coarse spatial scales such as counties or regions, although we note the growth of 'lifestyle' databases collected by commercial companies such as NDL and ICD. In these cases companies are generating huge quantities of individual household information from questionnaires and product registration cards. Although the responses are not a systematic sample, the sheer volume of responses suggest that lifestyle databases are becoming a valuable addition to Census data (see also Birkin 1995).

Modelling methods for linking these data sets, along with transition-rate data generated by OPCS for births, deaths, marriage, and so on to allow for an updating of Census and related information clearly have an important role to play. Methods for population updating at fairly coarse spatial scales have been available for a long time (Rees and Wilson 1977), but techniques that allow for the updating of both household and individual attributes, extending beyond age and sex categories, are in their infancy. Again, microsimulation methods offer much promise, and the shell of a systematic microlevel updating procedure has been constructed by Rees et al (1987) – see also Section 7.5.6.

3.4.4 Forecasting

Whilst updating takes us from the past to the present, requirements to examine how change may take place in the future is important in a number of contexts. For example, health, social services, and education authorities need not only to know what the future structure of the population is likely to be but also to have a reasonably clear indication of the way this structure varies across space. The same requirement applies in other sectors, such as retailing and financial services.

Any type of forecasting will have to be based on some kind of model – even if it is only trend extrapolation. The best-developed methods commonly used for forecasting, such as econometrics, have paid relatively little attention to spatial analysis, partly through the lack of appropriately referenced data sets (nicely illustrated by Bailly et al 1992). Many of the methods used in updating, described in the previous section, can be applied to forecasting, particularly in the demographic area. The outputs of economic and demographic forecasts could then be used as the inputs to spatial models to develop a complete forecasting system – though these have been developed only rarely in practice.

3.4.5 Impact analysis

One of the most popular application areas for spatial modelling has been in assessing the impacts of new plans or proposals on the existing situation. This 'what if' simulation has proved valuable in a number of areas in both public and private sectors. (This is, of course, a kind of forecasting, but usually only short-term, concerned with the immediate impact of a new facility.) The development and location of any new facility involves large capital investments. It is therefore appropriate to attempt to estimate not only the likely revenue that will accrue on that investment but also the impact that the new facility will have on existing outlets. Again, as we saw in Section 3.2, GIS offers a methodology for estimating the revenue from new sites by buffering a likely catchment area around the new site and then overlaying the population within the buffer to estimate possible spending power in the catchment area. The difficulty arises over the allocation of that expenditure to the new store, and the estimation of likely impacts on existing stores.

On the other hand, spatial interaction modelling is ideally suited to this problem. A study area is divided into a set of residential zones, say postal sectors, and all centres and outlets are identified. The model is calibrated on existing flow data, after which the new outlet is introduced and its impact assessed. Although these models have been around for 30 years, only recently, as better spatial data and computer hardware/software became available, has their full potential been realized (Clarke and Wilson 1987). Again, this emphasizes the strengths that can arise from an integration of data systems and models. The chapters that follow provide many examples of impact analysis.

3.4.6 Optimization

We saw in Section 3.2 that one of the uses that is commonly made of a GIS is to find locations in a region where several criteria are met, such as, say, flat land, proximity to a motorway access point, and areas with a catchment population greater than 100 000. A great deal of effort has been applied to generating software for solving this type of problem. In spatial modelling the problem has been addressed from a different approach, that of optimization (Wilson et al 1981).

Optimization methods such as linear programming attempt to find the best solution to a stated problem, subject to a number of constraints being satisfied. They first appeared in the geographical literature through methods such as the travelling salesman problem and location-allocation modelling. Nowadays a much larger class of methods exist, particularly in the non-linear programming area. The great advantage of these methods over standard GIS systems is the way in which they can handle spatial interaction. For example, in site location problems they can consider a particular solution (say, 50 particular sites chosen from 500 possible ones) and calculate catchment populations and potential revenues very easily. Existing GIS can help possibly in identifying local sites; certainly a marriage of methods will prove invaluable. Examples of optimization will be given in Chapters 5 and 9.

3.5 CONCLUDING COMMENTS

Section 3.4 presented illustrations of the general way in which models can contribute to the analysis of location problems, and of their considerable advantages over the spatial analysis routines in proprietary GIS. It is important to appreciate at this stage that there is no shortage of model-based packages available on the market, particularly related to retailing. Kohsaka (1993) gives a useful checklist. However, what is increasingly clear is the need for these to be linked not only to a wider range of analytical procedures but also to a computer graphic/GIS environment (see also Batty 1992). Hence, we have argued that the link should be made between currently conceived GIS and state-of-the-art spatial modelling. We believe that model-based analysis provides the central focus for GIS in the social sciences (it may also be true in the physical sciences) that will enable them to provide the types of information that prospective users require. Hence, two crucial points have emerged from the argument. First, GIS development is important for the storage, manipulation, and presentation of much geographic data. Second, model development is vital for generating more useful information in a variety of decision-making and planning environments.

It is important to end with a re-emphasis on the importance of the planning environment (outlined in Chapter 2). In most organizations planning involves a number of distinct stages that should take the form of a planning cycle. The first stage may involve a detailed descriptive analysis of the organization's products, its structure, its competition, and its performance in the market at a number of scales *vis-à-vis* its competitors. Following the setting of objectives (for growth in market share, profitability, etc.), a number of alternative approaches will have to be identified for achieving these. The next step will be to evaluate the alternatives using a number of criteria (financial return, risk, etc.) and make a set of recommendations. Once the plan is implemented, a monitoring process should be established to ensure that the objectives are indeed being met, and, if not, that remedial action is instigated. Whilst recognizing in practice that the planning process can be more complex than this, this broad set of stages is probably common to most companies in the retail and related sectors.

The problems faced in undertaking this process are typically ones of information availability and access to tools that will identify and evaluate a range of scenarios. The argument in this chapter is that most of what is available in the current GIS industry is limited in use to descriptive analysis, and far too restricted in real analytical power. What is required is the suite of model-based tools described above in conjunction with data storage, retrieval, and graphic facilities. Only by examining in detail the impacts of alternative scenarios for growth will we help to improve business planning (of whatever types) and performance.

CHAPTER 4

Spatial analysis in retail site location

4.1 INTRODUCTION

The use of GIS and spatial modelling in applied retail store location analysis is a relatively new phenomenon, despite the latter having a long history in the academic literature. It has come about through greater access to both spatial data (increasingly in digital form) and new packages designed to solve site location problems (proprietary and customized packages – see Longley and Clarke 1995). However, site location analysis itself has a much longer history, dating back to the pioneering works of Applebaum and Huff in the 1960s. Since those days we have witnessed the promotion of a wide variety of different methodologies and appraisal techniques. It is recognized that it is important to place modern methods in an historical context if we are to argue that significant progress has been made. Hence, the aim of this chapter is to review this progression from relatively unsophisticated approaches to the analytic frameworks now provided by spatial modelling.

In addition, we would like to emphasize the broader importance of geography within retail planning and marketing, over and above site location analysis itself. Hence, the chapter begins with a brief review of why a geographical perspective is important in understanding retail growth and change.

4.2 RETAIL GROWTH

The battle for retail market share during the past two decades has been fierce and extremely competitive. Firms have become larger in the wake of a variety of corporate strategies aimed at eliminating the power base of both actual and potential competitors. The net result of such actions has been the increased market shares of the leading retailers in all sectors of the retail economy. Wrigley (1993) suggests the market share of the top five grocery retailers in the UK increased from under 25 per cent in 1982 to over 60 per cent by 1990. The power of certain retailers is now on a par with that of multinationals in other sectors of the economy. Marks & Spencer alone claim to have 33 per cent of UK lingerie sales and 25 per cent of the menswear market.

Retail growth is achieved through a variety of business strategies, and many authors highlight the importance of understanding the corporate *retail marketing mix* if we are to understand the growth of individual companies. According to McGoldrick (1990) and Walters and White (1987) there are many components to this 'mix', and they illustrate the complex interplay of many economic and market forces. However, we can narrow this mix down to the four popular Ps of marketing: product, price, promotion and place. In briefly discussing each of these we set the context for our concentration on store location techniques and the development of a SDSS for the retail industry (Chapter 5).

4.2.1 Product

The first element is the basic building block: a company must have a product that the market requires. The first retail multiple groups to develop in the nineteenth century concentrated on the staple products of food, clothes, shoes, medicines, and newspapers/books (the last with the advent of the railways, which provided both an outlet space for the retailing of newspapers and a distribution system to supply those outlets across the UK). The most profitable companies over time have been those best able to keep abreast of changes in consumer tastes, habits, and behaviour and to modify their product range accordingly. Some of the most successful retailers of recent times have joined the market with new products (and in many cases methods of retailing) as they have recognized gaps or niches unfilled by the existing major retailers. Sir Terence Conran's vast Storehouse empire in the UK began with a recognition that furniture retailing in the 1950s and 1960s was drab and old-fashioned, and that it failed to meet the needs of the new market of younger families looking to spend greater disposable incomes on more exciting designs.

A similar story existed in the fashion industry with the new designs of Mary Quant and Laura Ashley (Kay 1987). These retailers concentrated on so-called 'focused strategies' of growth, identifying a particular type of consumer (young, old, low-income, high-income) and targeting this market through specialized merchandise and trading styles (see Dibb and Simkin 1991). The latter has reached new levels of sophistication with groups such as Laura Ashley, a name now synonymous with a special lifestyle. Indeed, other companies are able to plug into that lifestyle idea – a recent Toyota brochure has a Corolla hatchback car parked alongside a Laura Ashley store, since these shops are popular with the higher-income groups in the UK (see Figure 4.1). Other recent examples of niche retailing can be found in all sectors of the retail market: food (in the shape of freezer stores and more recently fast-food outlets), clothing (Tie Rack, Sock Shop), cosmetics (Body Shop), and 'leisure' in its broadest sense (Early Learning Centre, Kwik-Fit, Disney Store).

Figure 4.1 Brand marketing through the lifestyle concept
(*Source*: Toyota (GB) Ltd)

For many companies, an alternative to product development *per se* is to diversify the range of products in their portfolio. The idea is not new, of course: the grand department stores of the nineteenth century built on the desire of consumers to browse through a wide range of products in one store and enjoy the unique experience of shopping under one roof. Such *scrambled merchandizing* remained the basis for the success of some of the most important multiple retailers of the mid-twentieth century, such as Woolworths (imported from America) and Boots (broadening its range of products from its base in drugs). More recently, product diversification has hit the most traditional of retailers: supermarkets have diversified into petrol retailing, whilst petrol retailers have begun to sell groceries.

As retail companies grow in size so their buying power also increases, to the point where they may be able to enjoy monopoly power over suppliers to make them provide unique products to that company. It is argued that this shift in power from manufacturers to retailers offers more potential for profits to the retailer than small increases in sales at the consumer end. Understanding the changing relationships between retailers and manufacturers is a crucial aspect of understanding retail growth, and sets the agenda for an important and neglected area of economic geography (see Wrigley and Lowe 1995).

4.2.2 Price

Price has always been important to retailers, and all companies will carefully judge the price their market can sustain. The first multiple retailers were able to build up scale economies and pass on these cheaper costs to the consumer in the form of cheaper prices (see Jefferys 1954). The current large supermarket retailers in the UK were really only able to enjoy substantial grow when resale price maintenance was abolished for foodstuffs in 1964. They were then able to put substantial pressure on the small corner shop by offering heavily discounted prices (Dawson and Kirby 1977, 1979). Today, certain grocery retailers have retained low price as their major growth strategy. Kwik Save, with its 'no frills retailing' message, has continued to obey the old adage of 'pile it high, sell it cheap' once synonymous with Tesco (Corina 1971). Similarly, new entrants to the UK grocery market from Europe and America are concentrating on this segment of the market. The latter, such as Costco, are operating from very large retail warehouse sites and aim to cut grocery bills by 25 per cent. These low-cost leadership strategies are normally pursued in a large market segment of price-sensitive customers, where high-volume sales of a limited product range are the norm.

Pricing policies are related to geography in a number of ways. First, retailers may recognize the differential spending power of different regions (caused by differential regional wages and incomes) and adopt flexible pricing policies which may offer lower prices to markets perceived as being less affluent (see McGoldrick 1988). Secondly, and more subtly, the differential spending power within regions and, indeed, towns and cities may or may not attract discount retailers to locate there. Wrigley (1993) shows the distribution of Aldi stores in the UK, and the concentration of outlets in the less affluent heartland of the West Midlands and the north of England (see Figure 4.2).

4.2.3 Promotion

Another key factor in explaining the rapid rise of many multiple retailers has been advertising and company/product promotion. These clearly take many forms. Direct advertising through the media of newspapers, magazines, TV, and bill-boards is the most obvious form of company or product promotion. A *Which Report* in 1989 estimated the national advertising budgets of the major car manufacturers in the UK as follows:

Ford   £38m
Rover   £38m
VW   £29m
Renault   £23m

For Ford this works out at an average of £300 per new retail sale. Added to this total is further expenditure on dealer advertising, dealer support, and other promotional activity such as sport sponsorship. Through national or regional advertising campaigns, companies are able to relay a certain image

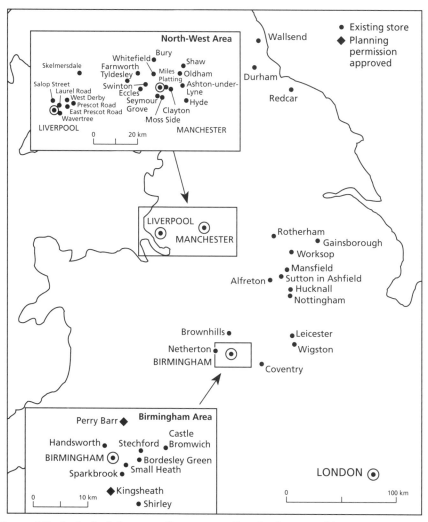

**Figure 4.2 Arrival of the new discount retailers in the UK: Aldi's UK store
        network 1992**

(*Source*: Wrigley 1993)

to the consumer which can then be strengthened through local store circumstances. The Kwik Save customer gets what she or he has been promised – no frills in exchange for cheaper prices. The Sainsbury's or Marks & Spencer customer is looking for something beyond simply the cheapest prices: she or he is looking for quality and high levels of choice and service.

It is important that the facilities offered by the retailer match these images that they themselves have created. As we saw in Section 4.2.1, much of the retail revolution of the 1960s caused by product and market segmentation was also driven by a desire to improve retail store images. Design is now a key word in modern retailing, and we have witnessed a huge growth in the design industry paralleling the growth of companies like Habitat, Laura Ashley, and British Home Stores (note the change in the latter's 'BhS' logo to a small, italicized *h* to reflect the introduction of 'colour, style and flair' – West 1988). The importance of design to retail growth is perhaps best exemplified by Next. Their phenomenal growth in the UK during the 1980s was based on a clearly defined targeted market (fashion for the 25–40 year-old market) and an internal store design which allowed plenty of open space and rest space to encourage a feeling of shopping as a pleasure not a chore.

With the increasing sophistication of retail products and markets has come the rise of the retail marketing industry – the ability to attract more customers to the marketing mix on offer. Despite the widespread acknowledgement of the four Ps of marketing, 'place' is often planned with little understanding of the importance of space. Although increasing product or company awareness is crucially important, it is wasted expenditure if the consumer cannot translate that awareness into a purchase because there are no stores or outlets in the vicinity. The most extreme example is probably the insurance industry, where millions of pounds are spent on national advertising and sport sponsorship. However, it has been argued (Clarke 1993) that whilst this raises company awareness it does little for additional sales, since consumers cannot purchase policies in unique company stores and, in any case, sales are controlled by brokers.

4.2.4 Place

The importance of the link between location and sales means that the idea of place is central to successful growth strategies. The elements introduced above all have a spatial component. In terms of product, companies may wish to vary their product mix from place to place (for example, Toyota will look for different distribution strategies in relation to Starlets, at the bottom of their range, and Lexus, at the top). We have already discussed the explicit role of space in pricing policies. In terms of promotion, issues of local and regional advertising are crucial to the successful development of a product in the market.

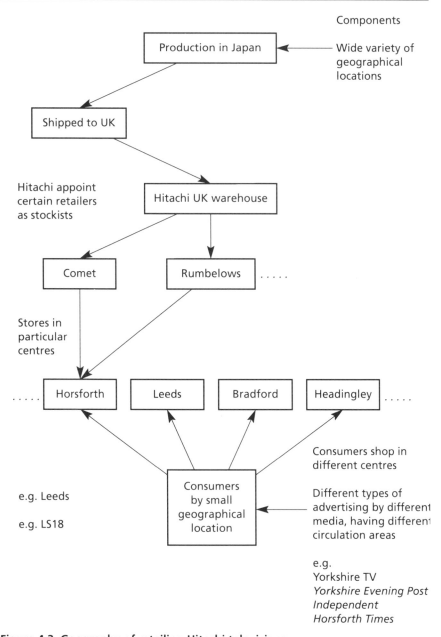

Figure 4.3 Geography of retailing Hitachi televisions

However, the most important element of place is the organization of the retail branch network to increase market penetrations in as many regional markets as the firm perceives as important. Figure 4.3 shows how geography plays a part in the marketing and distribution of a typical product, in this case Hitachi televisions. A company such as Hitachi must assign distributors to handle its goods in a foreign country. The chosen distributors are unlikely to have a uniform geographical presence across the country; hence, through its choice of distributors alone, Hitachi will have an uneven market share across geographical markets.

Geographical growth can be brought about through a number of diverse strategies. A first major way of obtaining geographical growth is through mergers and acquisitions. This is normally the fastest method of getting higher market penetrations, particularly in areas where an organization may be currently under-represented, and it often allows the direct strengthening of market share by the elimination of competition. There are many examples of this across all retail sectors. In the pub industry Tetleys, a leading UK brewer, had a firm no-merger policy in the mid-1950s, concentrating instead on organic growth. However, with the increasing pressure of competition, the company quickly realized it could gain strength and security through mergers and acquisitions. By 1960 Tetleys had expanded into Sheffield through the acquisition of Duncan Gilmours (350 pubs) and to Bradford through the acquisition of William Whittakers (119 pubs), and had further consolidated the Leeds market with the acquisition of Melbourne Breweries

Table 4.1 Expansion of Sears through merger and acquisition

1950s	J. Sear (shoe shop)
Mergers/acquisitions	
1953	Freeman Hardy and Willis
	Trueform
1954	Curtess
1956	Manfield
	Dolcis
1959	Garrard, Mappin and Webb
1962	Saxone
	Lilley and Skinner
1965	Selfridges
	Lewis's
1970	William Hill
1978	Olympus Sports
1985	Fosters
1986	Millets Leisure
1987	Hornes
	Zy
	Bertie

(245 pubs). Expansion across the Pennines into Lancashire came with the merger with Walkers of Warrington to produce the new organization of Tetley Walker. The latter then become part of the vast Allied-Lyons empire and has recently come under the international banner of Carlsberg-Tetley.

Sears, the shoe retailers, have a similar history of mergers and acquisitions aimed at eliminating competition and increasing market shares. Throughout the 1950s and 1960s they acquired increasing numbers of the famous British high street shoe-shop names (including Freeman Hardy and Willis, Trueform, Curtess, Manfields, Saxone, and Lilley and Skinner – see Table 4.1). More recent examples of important mergers in the UK can be seen in Table 4.2. The managerial advantages and disadvantages of mergers and acquisitions are discussed in detail in McGoldrick (1990) and Jones and Simmons (1990).

Whilst mergers and acquisitions were common features of the battle for the high street during the 1970s and 1980s (Kay 1987), most organizations also grew through organic development. This involves the company adding new stores to its own network, particularly to secure growth in new regions of the country. The history of the major multiples in the UK illustrates the importance of this method of growth. The first Boots store was opened in Nottingham in 1877. By 1914 Boots had some 550 branches around the country. More recently, the first Body Shop appeared in Brighton in 1983. By 1988 the chain had over 500 outlets. Perhaps because of such rapid growth, many supposedly national retailers end up with a very unbalanced retail network and huge regional variations in store outlets per head of population. This is shown by Langston et al (1995) for the major UK grocery retailers: Figure 4.4 plots the regional market shares of three leading UK grocery retailers. This means that new opportunities are always possible, provided also that the distribution network can support such growth (the importance of the location of distribution centres and regional growth is explored in Sparks 1986 and Fernie 1990).

Table 4.2 The merger mania of the 1980s

1984	Ward White	– Halfords
	Dixons	– Currys
1985	Ward White	– Owen Owen
	Sears	– Fosters
	Burton	– Debenhams
1986	Ward White	– Payless
	Ratners	– Samuels
	Next	– Grattan
	Dee Corp.	– Finefare
1987	Argyll	– Safeway

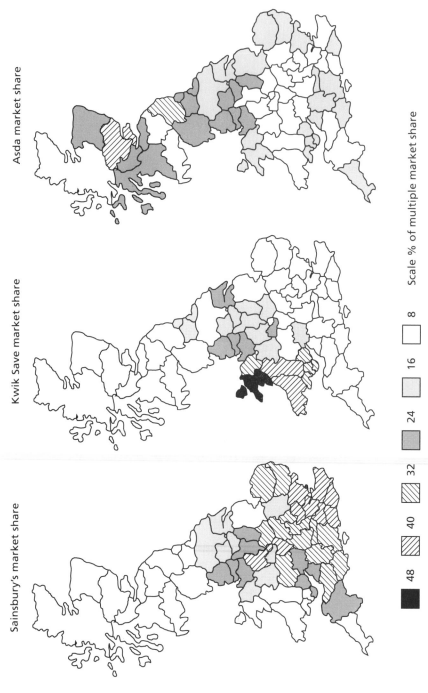

Asda market share

Kwik Save market share

Sainsbury's market share

Scale % of multiple market share

48 40 32 24 16 8

Figure 4.4 Examples of regional market penetrations in the UK grocery market

Franchising is a form of retail growth which involves collaboration between multiple retailers and independents. There are some notable success stories achieved through such partnerships. The expansion of Body Shop stores, mentioned above, has been achieved through the franchise system, where each store offers the same trading format and products but local management is left in independent hands. There are advantages to the main corporate group in terms of fewer financial investments and the ability to concentrate on product development rather than store management. The franchisee has the advantage of sharing in profits (which gives more incentive for success than 'tied' management) and the comfort of a tried and tested retail format. Further discussion on this form of retailing appears in McGoldrick (1990) and through various articles in *IJRDM*, the *International Journal of Retail and Distribution Management* (1991).

4.2.5 Summary

The retail industry is driven by a subtle mix of all the factors described above. This emphasizes the importance of getting the right balance in the marketing mix, and of making sure that growth strategies are supported by a store network of the right size and type and in the right location. The senior management team must formulate a strategy for growth based on these factors and market it successfully to the public. This strategy may change over time. Tesco, in the UK grocery industry, is a good example. Corina (1971) gives the historical account of Tesco's rise to power, founded on the principle 'pile it high, sell it cheap'. During the 1970s and 1980s the company changed its profile, aiming to rival Sainsbury at the upper end of the quality market. A new store location team was put into place to find sites that would enable Tesco to compete in this market. The design of Tesco's stores was upgraded and the product range extended into more quality items. There was a concern that Tesco had got the mix wrong, or at least had concentrated too much on the up-market image and have abandoned their traditional customer base too readily (Connon 1993). However, trading performances in the mid-1990s suggested that perhaps the strategy was right after all, since Tesco overtook Sainsbury as the leading grocery retailer on market share for the first time in the UK.

4.3 STORE LOCATION RESEARCH

4.3.1 Introduction

Retail companies have clearly practised some form of store location research from the first days of expansion in the nineteenth century. It is likely that such research would have been simple, with most multiples building networks based on regional or national hierarchies of urban centres. Indeed, for many new groups emerging on the high street this would still be a sensible tactic. However, as we have seen above, retailing has become very much more sophisticated since the 1960s as markets have become more segmented (and hence there is the need to find the right sorts of customer), consumer behaviour more varied, and the battle for market share much fiercer. In addition, the growth of suburban and out-of-town developments has made location decisions harder to make. As the Tesco chairman Ian Maclaurin emphasizes: 'There are only a finite number of superstore sites available in the UK. . . . It is my job to make sure that we get our fair share of the remaining sites' (quoted in Penny and Broom 1988: 107).

The academic community became interested in store location research in the 1960s, particularly in the US, where the movements out of town came first (see Applebaum 1965, Huff 1963). The main aim was to evaluate sites to determine the potential store sales and hence the probability of success. In order to accomplish this, the strategy had to assimilate information about potential sites and their existing levels of both demand and competition, in conjunction with corporate concerns such as return on investment, trading policy, and company character. UK summaries of such research appeared later in the literature (Cox 1968, Davies 1977, Davies and Rogers 1984).

At the outset, it is important to emphasize that store location research is not simply about appraising sites for new store openings. Moore and Attewell (1991: 21) summarize the Tesco philosophy:

> From the relatively narrow confines of site location analysis the company now calls upon the Site Research Unit to perform the broader function of Market Analysis whenever a locational element is involved. By this we mean the ability to advise not only on new store potential but on the past and future performance of existing branches, the purchasing behaviour and preferences of their catchments and the trading strength of other retailers' branches.

Beaumont (1991a, b, c) gives a checklist of spatial questions associated with new stores or existing stores. From this we might pick out the following key geographical concerns:

- Who:
  - are our customers?
  - should our customers be?
- Where:
  - are our customers?
  - should we develop to capture new customers?
  - are our competitors?
  - are our competitors developing?
  - should we advertise?

It is useful to review the wide range of techniques on offer to the retail analyst. Before proceeding, however, it is important to say something about geographical resolution. Ghosh and McLafferty (1987) provide a three-fold spatial classification which is useful for classifying locational decisions. At the broadest level there is usually a decision to move in to a region of the country (say, Florida in the US or south Wales in the UK). They term this 'market selection'. At the next level of their spatial hierarchy there are two further decisions on where to locate within these broad regions. Is Cardiff preferable to Swansea or are both good sites? If Cardiff is a good site, then where in the Cardiff region should that new store ideally be? Ghosh and McLafferty label this 'areal analysis', and it is these two intermediate levels which most concern us in the rest of this chapter. That is not to say that Ghosh and McLafferty's final spatial level – the site itself – is unimportant. Indeed, once a region within a town is chosen we recognize that it is crucial to find the right site on the ground – at a junction, close to a bus station, etc. For example, Epstein (1971) refers to merchants and site selectors who claim 'magical powers' for being on the 'homeward bound' side of a main road. Bennison and Davies (1980) and Brown (1992) give a long checklist of micro-factors which are important for site location within a shopping centre. These micro-site attributes are difficult to incorporate into aggregate models (although see Section 4.3.6.1) and are probably best considered by retail experts on site visits. Figure 4.5 summarizes the three levels of spatial analysis for the site selection.

4.3.2 Trial and error

This approach is usually thought of as the simplest in terms of spatial analysis. It normally involves the on-site decision of a senior member of staff who gets a 'gut feeling' for a location by walking the streets (sometimes referred to as 'counting the chimney stacks'). As Davies (1977) points out, this should not be belittled since these individuals will usually have the ability to offer very good instinctive judgements. Simkin (1990) emphasizes how important gut feeling/intuition still is in the UK amongst a variety of retailers (see Table 4.3). In addition, the importance of site visits does not diminish even when more sophisticated techniques are on offer. Towsey (1972: 40) rightly points out that locational research should be a mixture of 'science and old-fashioned nose'.

However, there are a number of obvious drawbacks with such an approach. First, it will remain highly subjective and depends entirely on the experience of those making such decisions (there are a number of boardroom anecdotes of senior staff all having to write their own estimates of a new store's potential to see how much agreement on a site exists). If that decision is in the hands of one individual then there is always the problem of that individual leaving the company. Second, it is a very time-consuming

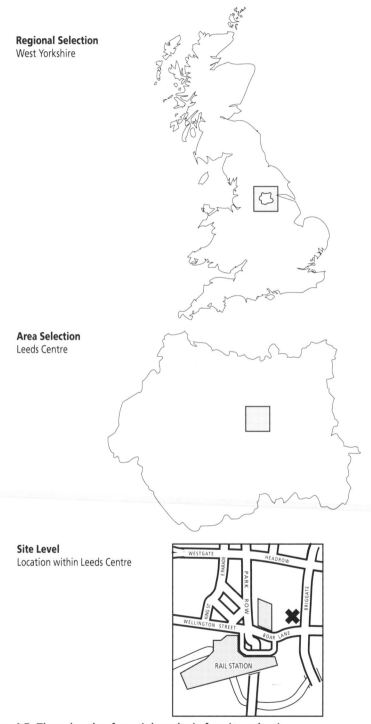

Figure 4.5 Three levels of spatial analysis for site selection

and expensive exercise. For those organizations with planned expansion programmes it may not be logistically feasible to visit all possible sites. In the US, for example, the fast-food retailer Taco Bell plans 3000 new stores over the next five years.

Even though many senior retailers will openly declare they have never made an unprofitable decision through such gut feelings, the market is littered with failed examples and store closures. Even profitable sites are open to investigation – could greater profits have been made elsewhere? Moreover, the increasing complexity of the retail scene makes it harder to make such gut feelings work. It is difficult even for the most experienced senior executive to stand on a green-field site and predict the drawing power and revenues which might accrue to a new store.

4.3.3 Analogue techniques

Analogue techniques are very common procedures for site location in the UK and US, even with those companies having large site location departments such as Tesco and Sainsbury's. The basic approach involves attempts to forecast the potential sales of a new (or existing) store by drawing comparisons (or analogies) with other stores in the corporate chain that are alike in physical, locational, and trade-area circumstances. This may be done 'manually' or through regression techniques (see Section 4.3.4). If you are evaluating a new store site in, say, Harrogate can you find an existing store location around the UK that has the same (or similar) population and trading characteristics as Harrogate? If so, you can attempt to draw analogies with the trading performance of the store in that other town. Alternatively, the procedure may work by trying to find sites which are analogous with the top-performing stores within the company. That is, if Stroud is performing very strongly, can the analyst find sites elsewhere in the country which match the characteristics of the Stroud site?

Table 4.3 Survey of retailer's assessment techniques (*Source*: adapted from Simkins 1990)

Company type	Most common technique
Department stores	Checklist
Variety stores	Checklist and analogue
Out-of-town warehouses	Analogue
Grocery superstores	Analogue and regression
High street multiples	Gut feeling/Analogue
Small multiples	Gut feeling
Financial outlets	Gut feeling

The success of this approach depends on whether or not you can find similar sites across the country and whether you believe you can successfully transfer trading characteristics across geographical locations. This again depends on the experience of the location analyst and his or her team. According to Moore and Attewell (1991: 24) Tesco are improving in this area: '[The] greater understanding of the way in which existing stores trade has been fed back into the sales-forecasting process through an increased appreciation of analogue store performance.'

Apart from the required experience, a second problem remains the variable performance of stores across similar geographical markets. In reality, a wide variation in performance is frequently found between outlets in a retail chain. If a similar geographical catchment is found for the new store, what happens if the analogous store is currently over- or under-performing? Thirdly, it is extremely difficult to evaluate green-field sites in this way. These may have catchment areas greatly distorted by local transport networks, and it might prove impossible to import revenue predictions from other towns or cities.

A similar approach to the analogue method has been to follow the behaviour of other (larger) retailers and base store location decisions on whatever decisions they make. This has been labelled the *parasitic approach*. In the early days of the British high street, many new multiple groups would simply follow Marks & Spencer, Woolworths and Boots to new locations. The practice is still common today, especially for smaller retailers. In the US, for example, Mason and Meyer (1981) quote the (then) strategy of County Seat: 'If a Penney, Sears, Wards or local department store is going to go there then they have already done demographic studies. It almost sounds too simple but that really is our strategy.'

4.3.4 Regression techniques

The most commonly used statistical technique for estimating store revenues is the multiple regression model, which builds on the philosophy of the analogue procedure (see Rogers and Green 1979). Regression analysis works by defining a dependent variable such as store turnover and correlating this with a set of independent or explanatory variables. Coefficients are calculated to weight the importance of each independent variable in explaining the variation in the set of dependent variables. The model can be written as:

$$Y_i = a + b_1 X_{1i} + b_2 X_{2i} + b_3 X_{3i} + \ldots + b_m X_{mi} \qquad (4.1)$$

where:

Y_i is turnover (the dependent variable) of store i;

X_{mi} are independent variables;

b_m are regression coefficients estimated by calibrating against existing stores;

a is the intercept term.

Fenwick (1978) gives an example of these variables for a building society. Keeping the above terminology:

X_{1i} = Average age of persons in catchment area of branch i;

X_{2i} = Average socio-economic status in catchment area of branch i;

X_{3i} = Number of years branch i has been established;

X_{4i} = Number of new houses under construction in catchment area of branch i;

X_{5i} = Total number of building societies in catchment area of branch i.

Although the models allow greater sophistication and objectivity, there remain a number of problems. The primary weakness of such models is that they evaluate sites in isolation, without considering the full impacts of the competition or the company's own global network. As the above building society example shows, the level of competition is incorporated by the simple absence or presence of stores. A second major weakness is the problem of heterogeneity of sample stores. This was also seen as a problem with analogue techniques. That is, how easy is it to find a sample of stores which have similar trading characteristics and catchment areas (see Ghosh and McLafferty 1987)?

A third problem relates to the basic feature of regression analysis which assumes that the explanatory variables in the models (X_{mi}) be independent of each other and uncorrelated. In many retail applications this is not the case – independent variables such as floor-space and car-parking spaces may be strongly correlated. This can lead to unreliable parameter estimates and severe problems of interpretation. The so-called multi-collinearity problem has received much attention in the literature (Lord and Lynds 1981, Ghosh and McLafferty 1987). However, through careful analysis and interpretation many of these problems can be overcome. Most poor applications of multiple regression in retail analysis have shown statistical nativity and limited understanding of retail process.

Fourthly, and from our point of view most importantly, there is the limitation that regression models fail to handle spatial interactions or customer flows adequately. That is, they do not model the processes (spatial interactions) that generate the flows of revenue between residential or workplace areas and retail outlets. Although regression models may sometimes demonstrate impressive descriptive powers (through their ability to reproduce the variation in sales across a network), the absence of any process modelling leaves us sceptical as to their ability to undertake impact analysis with any confidence.

4.3.5 Catchment area analysis

This is a broad set of procedures aimed to measure more objectively the size of existing centres (and hence their potential) and to understand the breakdown of their catchment areas in terms of population structure. The size or importance of a centre can be measured relatively easily by standard floorspace statistics available from local authorities or private sector organizations such as Town Centre Plus (with its Focus product), GOAD, and the Unit of Retail Planning and Information (see also Schiller and Jarrett 1985). Such aggregate statistics can be broken down by type of retailer present in order to gauge the 'quality' of that centre. In effect, the procedure is concerned with compiling as much information as possible concerning the centre and its neighbourhood. For this reason it is also known as 'the checklist procedure'.

Once a measure of centre importance has been estimated (and the centre matches certain criteria concerning corporate views on the minimum size of centre needed for success), the analyst then looks towards estimating the catchment area of the new store. The most popular approach is to base estimates on 'drive time bands'. From existing stores in the chain it would be possible to undertake questionnaires to determine how far people generally travel to visit the store and the magnitude of the distance decay effect from that store. These analogies can be translated to the new store area and drive/travel time bands calculated around it. Then it is possible to calculate the resident or workplace population within each of the drive time bands.

Such catchment area studies may be enhanced by the use of geodemographic systems, which attempt to profile catchment areas into customer segment types. This is especially useful for those retailers who have customers concentrated in certain geodemographic segments and are keen to find localities of the 'right type' for their products. Geodemographic systems in the UK have been available since the late 1970s (CACI's ACORN system being the earliest commercial application in the UK) and have proliferated since the publication of the 1981 Census. These small-area profiles are based on a multivariate analysis of a large number of variables associated with small areas. This produces a limited number of single-dimensional classifications of neighbourhoods such as enumeration districts. Small areas that fall within the same cluster classification can be considered to be alike and to contain similar types of households. Figure 4.6 shows the ACORN classification mapped by Hirschfield et al (1993) in St Helens, near Liverpool in the UK. For a detailed history and review of these techniques see Beaumont (1991c) and Brown (1991).

The importance of geodemographic systems should not be underestimated. As companies increasingly target their stores at different members of society, these methods will remain popular. Sears, for example, own many different shoe shops in the UK which trade under a variety of names targeted at different consumer groups. Dawson and Broadbridge

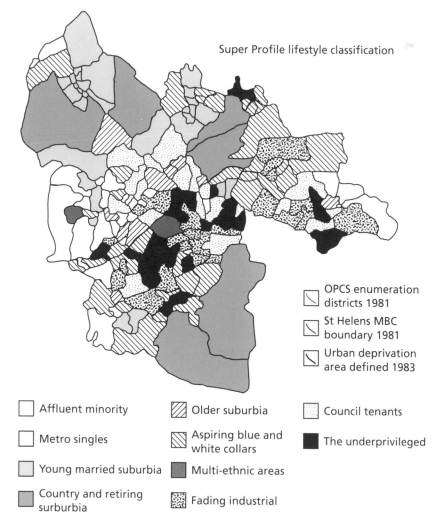

Super Profile lifestyle classification

⌑ OPCS enumeration
districts 1981

⌑ St Helens MBC
boundary 1981

⌑ Urban deprivation
area defined 1983

☐ Affluent minority   ▨ Older suburbia   ⬚ Council tenants

☐ Metro singles   ◲ Aspiring blue and white collars   ■ The underprivileged

▨ Young married suburbia   ▦ Multi-ethnic areas

▦ Country and retiring surburbia   ▨ Fading industrial

Figure 4.6 Mapping Super Profiles as indicators of deprivation
(*Source*: Hirschfield et al 1993: 223)

(1988) list some of these outlets and the specific market they are aimed at. These are shown in Table 4.4. Clearly, an important planning task is to work out which stores should be in which locations to maximize corporate returns on investment. This can be achieved, at least partially, by profiling the catchments of centres in such a way as to optimize the brand offering.

As more information is required on lifestyles, it is likely that these geo-demographic systems will become increasingly sophisticated through a linking of population demographics with lifestyle information obtained from companies such as NDLI and CMT (Openshaw 1989b, Birkin 1995).

Table 4.4 Target market of Sears' stores (*Source*: adapted from Dawson and Broadbridge 1988)

Operation	Sector	Target/position
Fosters	Menswear	Middle market, 15–30
Your Price	Menswear	Keen-priced fashion
Bradleys	Menswear	30–50, traditionalist
Hornes	Menswear	25–40, up-market
Zy/Jargon	Menswear	Younger men
Wallis	Womenswear	25–35 professional
Miss Selfridge	Womenswear	Young fashion
Curtess	Footwear	High-volume, low-price
Freeman Hardy and Willis/Trueform	Footwear	Family
Dolcis/Tiptoe/Bertie	Footwear	Fashion
Saxone/Manfield/ Lilley and Skinner	Footwear	Quality/high price

Lifestyle data is information collected about individual households, through the use of self-completed questionnaires. Since it is collected at this micro-level it avoids the problem encountered by using Census data, namely the fact that it is only available at an aggregate level. Using lifestyle data, it is possible to identify the number of households that have specific combinations of characteristics, such as 'Volvo-owning golfers with an interest in fine wine'. An early example of such a package is SHOPPiN, an amalgamation of Pinpoints' geodemographic PiN system and Nielsen's Homescan lifestyle consumer panel, which detailed the shopping habits of 100 000 households (Ody 1989). A second example is Equifax, with its geo-demographic system based on Census and lifestyle information (Sleight 1993). It is probably true that lifestyle data offer a more precise way of targeting particular customer groups, as well as being a powerful tool for direct marketing. Its main drawback, however, is that it is not a complete Census of the UK population and has under- and over-representation of certain groups. However, the largest lifestyle database (collated by NDLI) contains over 10 million households – nearly 50 per cent of the UK total. These systems are set to have a large impact on geodemographic marketing tools in the 1990s.

Geodemographics has attracted the attention of many retailers and has developed a keener interest in geographical analysis. Many users are now looking for more powerful tools and, as Openshaw (1989a) warns, it is important for geographers to build on this foundation of interest.

It is this sort of catchment-area analysis which has been greatly aided by the arrival of GIS packages. As we saw in Chapter 3, there are many examples in the literature of a GIS being used to buffer drive time bands around new stores and then to overlay population characteristics associated with those drive time bands (Howe 1991, Elliott 1991, Ireland 1994, and see Figure 3.6 again). The incorporation of geodemographics into GIS has been a welcome initiative. At least this is beginning to make such systems more suited to client needs. Good examples include Pinpoint's PiN profile, which has been incorporated into ARC/INFO so that the geodemographics of catchment areas can be readily plotted (Beaumont 1991b, Reynolds 1991), and FINPiN, which is tailored to the financial service market. CACI's Insite System has been specifically targeted at retail businesses wishing to match catchment-area profiles based on Census and geodemographics with those obtained from their customer databases. They are working with a number of high street retailers, including Norweb, Britdoc, Budgens, Woolworths, and Yorkshire Building Society (CACI 1993). Such bespoke geodemographic systems are also increasingly available within general GIS packages for other regions of Europe (Hinton and Wheeler 1992, Reynolds 1993). Indeed, CCN have recently launched a pan-European version of their popular MOSAIC system (Webber 1992, Birkin 1995).

The two principal drawbacks of these approaches are, first, the definition of the catchment area and, secondly, the treatment of the competition. The former is normally represented by distance or drive time bands, and it is often assumed that the store will capture trade uniformly in all directions. Even when drive time bands are drawn in relation to transport networks (Reynolds 1991), there is still the assumption of equal drawing power in all directions. These methods also give equal weight of importance to all households within a buffer. If a five-mile buffer is drawn around a new store as the primary catchment area then households close to the site are given the same weight (or probability of patronage) as those 4.9 miles away. In addition, the treatment of the competition is wholly inadequate. The presence of competitor stores will mean the real geographical catchment area of a new store will be highly skewed in certain directions. This can normally be shown in all appraisals of existing store catchment areas. Figure 4.7 shows the real catchment area of a grocery store in Elm Park, north-east London, which is highly skewed to the south and east due to the lack of competitor stores in either of these directions.

Similarly, there is no effective way in most GIS of estimating the new store revenue in light of the level of competition. As Beaumont (1991b) suggests, the method most often used is 'fair share', with the potential revenue of the catchment area being simply divided between all retailers on some *ad hoc* basis (type of retailer, level of floor space, etc.). Hence this methodology, whilst offering a useful overview of potential catchment-area revenue,

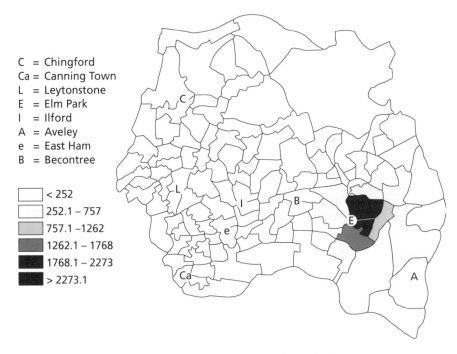

C = Chingford
Ca = Canning Town
L = Leytonstone
E = Elm Park
I = Ilford
A = Aveley
e = East Ham
B = Becontree

< 252
252.1 – 757
757.1 –1262
1262.1 – 1768
1768.1 – 2273
> 2273.1

Figure 4.7 Real shopping-centre catchment-area boundaries

is fundamentally flawed due to the inadequate treatment of spatial interactions and of competitor impacts. We revisit this issue in Chapter 5.

As we argued in Chapter 3, proprietary GIS are often too inflexible to handle the variety of real decision-making environments that the complex modern retail environment demands. It is timely to reconsider the words of one of the founder members of the store location research paradigm:

> To design store location strategy models one must be familiar with the business objectives and merchandising and operating policies of the firm for whom the task is being performed. Thus in a sense, model design is like tailoring a suit to fit the measurements of the man [sic] who will wear it.
>
> (*Source*: Applebaum 1965: 235)

Despite our major reservations concerning catchment-area analysis and GIS, we cannot deny that retailers continue to invest in such software developments and clearly find these of considerable benefit. Even Marks & Spencer, traditionally conservative in store location research, has now decided to invest in GIS technology and person power (Figure 4.8, and Ireland 1994). Other retailers are in turn bound to sit up and take notice of this initiative.

Whatever the take-up of GIS for the type of catchment-area analysis described above, we shall continue to argue that better methodologies are

St Michael

Geographic Information Systems

Central London   c. £30,000 + car + benefits

As part of a multi-million pound investment programme, Marks & Spencer is developing new stores in Britain and across the world. To ensure the optimum return on the programme, it is essential that the best possible locations are selected for the developments.

We now have an opportunity for a highly motivated professional to join our Site Assessment team, based in our Central London Head Office. You will be responsible for developing and implementing an integrated Geographic Information System for the Company.

We're looking for someone who thrives under pressure and possesses excellent analytical skills, including the use of statistical methods, and the

use of computers for applied geographical and market analysis. In-depth experience of one of the main GIS package is preferable, as is previous commercial experience.

In addition to a highly competitive salary, you will receive a non-contributory pension, life assurance and profit sharing after a qualifying period. Relocation assistance will be available where appropriate.

If you feel you can thrive in this demanding environment, please send your CV, quoting current salary to: The Recruitment Department, Marks & Spencer plc, Michael House, 57 Baker Street, London W1A 1DN.

We are an equal opportunities employer.

MARKS & SPENCER
HEAD OFFICE

Figure 4.8 Advertisement from 1992 showing Marks & Spencer's interest in GIS
(*Source*: Marks & Spencer 1992)

needed for handling the complexities of real-world consumer behaviour. We shall address this issue in the following sections.

4.3.6 Mathematical models

4.3.6.1 Spatial interaction models

For British academics, a remarkable feature of very recent developments in store turnover forecasting work has been the re-emergence of spatial interaction models, and the gravity model in particular, as proposed techniques.

(*Source*: Breheny 1988:76)

It is not surprising that many writers in retail geography have expressed such sentiments, given the relatively poor experience of spatial modelling applications in the public sector in the 1960s and 1970s. The lowest ebb was clearly reached in 1977, when a government warning was issued on the inappropriateness of the use of such methods in superstore public enquiries, mainly because of the seemingly conflicting evidence that such methods often produced (depending upon which party was actually producing the analysis: Department of the Environment 1977). Looking back today, it seems a harsh response to a situation caused more often by inadequacies in the use of the models than by the modelling techniques themselves. Indeed, many texts that have documented the decline in the use of spatial interaction modelling in retail planning have not ruled out their importance if applied in less cavalier fashion (Davies 1984). Another important caveat on that decision of 1977 is that the banning of models has not stopped the production of confusing and conflicting evidence at modern planning enquiries. This is very obvious from a report of a recent enquiry in Salford, near Manchester in the UK (see Bradshaw and Zalzala 1988). One cannot help feeling that a well-developed modelling appraisal, if undertaken by a neutral party, would help to add a more objective stance.

The resurgence identified by Breheny above is still in its infancy. Indeed, Simkin (1990: 33) observes: 'While mathematical models have been created, there is a dearth of operationally predictive models capable of reproducing meaningful and usable information for a company's management.' The situation is, however, changing rapidly. To illustrate the power of these models we first need briefly to explain their structure.

Let us label any residential zone such as a postal sector or enumeration district i and any facility location such as a centre or supermarket j. Then the number of people travelling between i and j can be labelled S_{ij}, and modelled using a spatial interaction approach:

$$S_{ij} = A_i \times O_i \times W_j \times f(c_{ij}) \tag{4.2}$$

where:

S_{ij} is the flow of people or money from residential area i to shopping centre j;
O_i is a measure of demand in area i;
W_j is a measure of the attractiveness of centre j;
c_{ij} is a measure of the cost of travel or distance between i and j.
A_i is a balancing factor which takes account of the competition and ensures that all demand is allocated to centres in the region. Formally it is written as:

$$A_i = 1/\sum_j W_j \times f(c_{ij}) \tag{4.3}$$

The model allocates flows of expenditure between origin and destination zones on the basis of two main hypotheses:

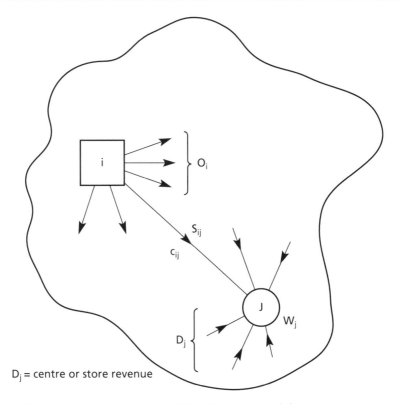

Figure 4.9 Pictorial representation of the shopping model

1 Flows between an origin and destination will be proportional to the relative attractiveness of that destination *vis-à-vis* all other competing destinations.
2 Flows between an origin and destination will be proportional to the relative accessibility of that destination *vis-à-vis* all other competing destinations.

The major variables of the model can be represented diagrammatically as in Figure 4.9.

The model works on the assumption that in general, when choosing between centres which are equally accessible, shoppers will show a preference for the more attractive centre. When centres are equally attractive, shoppers will show a preference for the more accessible centre. Note, however, that these preferences are not deterministic. Thus when choosing between equally accessible centres, shoppers will not always choose the most attractive. The models are therefore able to represent the stochastic nature of consumer behaviour. Neighbouring households would not be expected to behave in exactly the same way, even though their characteristics are similar. Equally, particular individuals and households will not always use the same retail centres.

Figure 4.10 **Customer flows to Bradford and Leeds from intervening postal districts**

Consider the example shown as Figure 4.10. Here we are looking at a string of postal sectors between the cities of Leeds and Bradford in the UK. To begin with, residents within postal sectors close to Leeds have very strong preferences for shopping in Leeds, as both the accessibility and attractiveness of Leeds as a centre dominate. Note that even in this situation we expect some flows to Bradford: there may be specialist shops, for example, which Leeds does not provide; or Leeds residents who work in Bradford may shop for some items from work rather than home. As we move towards Bradford, the flows gradually become more balanced before tilting the other way. Close to Bradford, the influence of Leeds fades, also to a small but non-zero level.

This simple example allows us to show an analogy between the spatial interaction model and Reilly's original 'breakpoint' model of retail attractiveness (cf. Foot 1981, Haynes and Fotheringham 1984). The breakpoint is a point of indifference between two centres, and in this case falls somewhere between BD11 and BD17 in Figure 4.10. In theory, the model can be derived in a number of ways. Wilson's (1967, 1970) derivation of the model using entropy-maximizing methods possibly remains the most significant, and shows that the model represents the distribution of flows which is statistically most likely in a given situation. The model can also be derived as an econometric (discrete choice) model in the logit style (cf. Domencich and McFadden 1975, Williams 1981) and is analogous to the intuitively pleasing but theoretically unfounded gravity-type models (for example, Huff 1963).

The models can be disaggregated in a number of ways. First, recognition of different types of consumer, such as car-owners and non-car-owners, is important in most real-world applications. Secondly, the destination attractiveness term can be disaggregated to include all sorts of centre or store attributes (Pacione 1974, Spencer 1978, Timmermans 1981, Wilson 1983). Thirdly, various forms of the distance deterrence term may be used and different transport modes introduced. Wilson (1983) provides a useful summary of the degree to which retail models can be disaggregated, whilst other authors have looked at new formulations of spatial interaction models which incorporate additional behavioural variables. Fotheringham (1983, 1986) has argued that the models need to be modified to allow stores in close proximity to other stores to have greater attractiveness to consumers. These competing-destination models measure relative accessibility of stores to one another to measure the degree to which stores located close to each other have a locational advantage over isolated outlets. This may be particularly important in comparison shopping. We shall introduce more disaggregated retail models in the applications discussed in Chapter 5.

The work on alternative parameter functions and the behavioural attributes of shopping-centre attractiveness has led to a new style of shopping model research, based on probabilities and utility concepts, leading to more formal statistical models related to discrete or stochastic choice theory of consumer behaviour. The most popular models are the multinomial logit model (Domencich and McFadden 1975) and the multinomial probit model (Miller and Lerman 1981). A typical logit model takes the following form:

$$P\,(d,\,m:\,dm_t) = \frac{\exp\,(u_{dmt})}{\displaystyle\sum_{dm}\exp\,(u_{dmt})} \tag{4.4}$$

where:

$P\,(d,\,m:\,dm_t)$ denotes the probability that individual t will choose destination d and mode m from a full set of alternatives open to him. (dm_t). u_{dmt} is the utility that individual t obtains from going to destination d by mode m and is assumed to be a function of the variables which describe the alternative shopping centres, the alternative modes and the socio-economic characteristics of individual t.

(Guy 1981:14)

On the face of it, these models seem to have quite a different form from the spatial interaction approach described above. However, it can be easily shown that the spatial interaction model is in principle equivalent to the logit model. The only substantive difference is in how various model parameters are estimated. The logit model typically relies upon individual-level data whereas the spatial interaction model can be run using more aggregate data.

The calibration procedures of mathematical models (that is, the procedure to set the model parameters to reproduce real-world interactions – see Section 5.5) are not problematic when good interaction data are available and when the models are not highly disaggregated. Many organizations are becoming 'data rich' and many will have at least some information on customer flows. If not, this is increasingly available from commercial agents or can be obtained using some kind of questionnaire or sample procedure. Once organizations realize the benefits of having such information, they are normally happy to spend resources to rectify their data deficiencies. The problem of calibrating highly disaggregate models is more of an issue, and one which the academic community continues to work on. A number of authors suggest transforming the interaction model into a probability model such as a Poisson regression model. Guy (1991), for example, concludes that this offers a more flexible approach to model estimation and can allow 'unusually complex' retail models to be calibrated with 'relative ease' (see also Cesario 1975, Flowerdew and Aitken 1982).

4.3.6.2 Location-allocation and optimization models

Location-allocation models are concerned with the location of facilities to serve the distribution of population best. Thus, they both locate facilities and allocate individuals to those facilities. Their interest is based on what is commonly known as the equity-efficiency problem. This recognizes that facilities need to be located to maximize their accessibility to individuals, yet at the same time that any network of facilities has to be efficient in relation to scarce resources. The best-known and most important of the location-allocation models is the 'p-median problem', which can be written (Scott 1971, Birkin and Wilson 1986) as:

$$\underset{\{S_j\}}{Minimize\ Z} = \sum_{j=1}^{p} \sum_{k=1}^{q} X_k h_{jk} c_{jk} \tag{4.5}$$

subject to:

$$\sum_{j} h_{jk} = 1 \tag{4.6}$$

$$h_{jk} = \begin{cases} 1 \\ 0 \end{cases} \tag{4.7}$$

where:

$S_j = (x_j, y_j)$ is a facility location;
X_k is the demand at k;
c_{jk} is the distance from j to k.

The problem is to find the optimal locations for p facilities (hence the name 'p-median') relative to q demand points or demand zones. (One of the nice features of this type of model is that they may equally be applied to demand surfaces in which the population is 'continuous', or divided into discrete zones.) In the basic model 'optimization' means transport-cost minimization, and the 'allocation' element of the problem comes through the zero-one variables $\{h_{jk}\}$ which allocate each demand point to the nearest facility.

The distinction between 'private' and 'public' sector location-allocation models has long been recognized (for example, Revelle et al 1970). Having said that, many of the best applications have been within the public sector, such as those of fire-station location planning (Richard et al 1990), school location planning (Church and Schoepfle 1993), and hospital location planning (Mayhew 1986, Rushton 1988), where distance to such facilities is all-important. However, one of the biggest drawbacks of traditional location-allocation models is that they take no notice of competition. As Ghosh and Craig (1984) point out, this may not be a drawback in public facility location problems but is a serious deficiency in the competitive environment of the private sector. They advocate a modelling framework to incorporate the competitive environment of retailing. The procedure works by allowing company A to locate in the best initial site, but then assuming that competitor B will engage in a similar decision-making process to determine the location of its own first site. The process continues until both retailers have located all their stores. This procedure is a step forward, since it recognizes that 'optimal' locations clearly vary once more than one store has been located (see also Ghosh and McLafferty 1987).

Location-allocation models are embedded within the latest versions of many GIS packages, such the Automatic Territory Assignment capability within Tactician (Tactics International 1993). The main application of this technology would appear to be for assigning sales areas and optimum base locations for a sales force.

An alternative approach is to embed spatial interaction models within a procedure which optimizes locations. There are two ways of going about this. One approach is to take something like the p-median and to replace the zero–one allocation variables $\{h_{jk}\}$ with a probabilistic model of market share, such as a spatial interaction or logit model. The resulting problem is called the 'p-choice model' by Ghosh and Harche (1993), and examples of its application include Hodgson (1978), Leonardi (1983), and Ghosh and McLafferty (1987). The second approach is to build location factors into spatial interaction models in the way suggested by Harris and Wilson (1978). These models have sometimes been referred to as models of spatial interaction, activity and structure (SIAS) (for example, Wilson and Birkin 1987). A comprehensive overview of these applications is provided by Wilson et al (1981).

The location-allocation models are mathematically quite flexible in the sense that the objective function (for instance, equation 4.5) and constraints (equation 4.6 and 4.7) can be varied to represent different kinds of situation. For example, it is possible to represent 'covering problems' in which the objective is to minimize the greatest distance which any consumer must travel to a facility (also known as 'minimax' problems, for obvious reasons); or 'plant location' problems, where the objective is to minimize the combined costs of production and distribution of a product. Some other variants are discussed in Ghosh and Harche (1993). Of course, one problem here is that packaged applications of location-allocation modelling (for example, within GIS systems) do not have this kind of flexibility. A second disadvantage is that it is rather easier to specify these models than it is to solve them computationally, as has been remarked elsewhere: 'location [-allocation] models are often extremely difficult to solve. ... Even some of the most basic models are computationally intractable for all but the smallest problem instances' (Church et al 1993: 1).

In the past, much the same has been true of the SIAS models, only probably more so. However the situation is changing. Birkin et al (1995) have shown how techniques from parallel computing can be used to help solve complex spatial optimization problems of this type. An application to the car market is described in Chapter 9 of this volume.

A broader criticism of location-allocation models is that retailers are rarely operating in geographical areas which have no existing outlets. New locations are likely to be sought in an environment which already contains many existing outlets. For these more common occurrences, spatial interaction models seem much more appropriate.

However, the search for optimal solutions may still be an interesting exercise, even though the majority of outlets are already located in the real world. It is useful, for example, to ask how far away our current network is from some sort of optimal configuration.

4.4 CONCLUDING COMMENTS

The extent to which retailers make use of the types of approach described above is an interesting question. From the experience of the authors, it varies from the relatively sophisticated to the embarrassingly naive. Whatever the current situation, retailing in the 1990s is ever-more characterized by increasing competitiveness, falling margins, and the exploitation of niche segments. Any competitive edge which a retailer can obtain in such an environment may prove to be precious. The pursuit of an optimal location strategy has the potential to provide such an edge. At the same time, the right sort of data is now being collected to facilitate better location decisions. Furthermore, there are dangers to decision makers in as much as the added complexity of the retail environment makes traditional methods of site assessment, especially those based on gut feel or intuition, progressively less effective. However, in most cases the methods which are being used to approach the problem of site assessment are simply not good enough to secure their users any advantage. The next step, in Chapter 5, is to outline the set of methods which, we believe, allow the available data to be mobilized as an effective aid to decision making.

Spatial decision support systems for retail planning

5.1 INTRODUCTION

In Chapter 4, we considered a variety of methods by which attempts have been made to link GIS and spatial analysis for retailers. Also, Section 4.3.6.1 showed, in outline, how to build a spatial interaction model of the retail process. In this chapter, we wish to show in more detail how to go about building a SDSS or IGIS around the construction of spatial interaction models of consumer behaviour. We will consider the sorts of data which are required before reviewing the types of application for these models within retailing.

The models are constructed from three key components: demand, supply, and spatial interaction. As in much conventional economic analysis, a lot of the interest revolves around the relationship between demand and supply. These two factors are considered in Sections 5.2 and 5.3. However, as the classical location theorists (amongst whom Christaller and Lösch are the most eminent) have emphasized, the introduction of geographic space adds a vital third dimension to this analysis. This spatial matching of consumers to providers is considered in Section 5.4.

Section 5.5 will introduce an appropriate set of mathematical models to represent these processes, and will discuss how these models may be calibrated to reflect a variety of situations. The interpretation and analysis of the rich and complex insights which the models offer will be considered in Sections 5.6 and 5.7. Throughout the chapter we shall be drawing on our experience from a variety of retail and quasi-retail sectors (see also Chapters 9 and 10).

The work undertaken by GMAP in the quasi-retail and financial services sectors is extremely diverse. Although specific applications for these different clients are highly customized, as we have shown elsewhere, it is nevertheless possible to describe the model structure adopted in general terms. The starting point is the type of model described in Chapter 4: the production-constrained spatial interaction model. However, as we discussed in Section 4.3.6.1, this model often requires extensive disaggregation in order to capture the variations in consumer behaviour for different retail activities. (Readers who experience difficulties with the notation or theory of these models are referred to Birkin and Clarke (1991) or Foot (1981) for a less technical introduction.) A typical level of disaggregation produces the following spatial interaction array:

$$S_{ij}^{kmn}$$

where:

i represents a small area (such as a postal sector, zip code or census tract);
j is a retail destination;
k is a person type;
m is a product type;
n is a retail organization or company.

A typical model structure is then:

$$S_{ij}^{kmn} = A_i^{kn} O_i^{km} \theta_j^{mn} W_j^m \exp(-\beta_i^k c_{ij}) \tag{5.1}$$

$$A_i^{kn} = 1/\sum_{jm} W_j^m \theta_j^{mn} \exp(-\beta_i^k c_{ij}) \tag{5.2}$$

where:

O_i^{km} represents the demand in small area i for products type m by people type k. The construction of this variable is described in more detail in Section 5.2.

W_j^m represents the 'attractiveness' of centre j for products type m; and θ_j^{mn} is the additional attractiveness of retailer n for product m at j. These variables are discussed in more detail in Section 5.3.

c_{ij} represents the travel time from residential places to centres. Travel times are discussed in Section 5.4, together with issues regarding the interaction data.

β_i^k is a distance deterrence parameter which controls the willingness or ability to travel. The calibration of this parameter is described in Section 5.5. (Note that this parameter could be disaggregated by m and n. Here we disaggregate by region i. In practice this would depend upon the quality of the interaction data – see Section 5.4.)

The application of GIS has sometimes been characterized as falling into four areas: 'what has been?', 'what is?', 'what if?', and 'what should be?' Our argument in relation to modelling would be that spatial analysis can enrich our understanding of each of these questions. Section 5.6 will consider the way in which the interaction variables can be manipulated to produce performance indicators for the 'what is?' questions. Section 5.7 will consider the manipulation of the model indicators to address 'what if?' and 'what should be?'

5.2 DEMAND

The concept of geographically differentiated markets is fundamental to retail analysis. Clearly, the reason that major retail activities are provided through extensive branch networks is in order to penetrate markets which are geographically dispersed. Similarly, for all but the most specialized of retail activities, small branch networks will only be suitable for the penetration of distinct, localized markets.

In classical location theory, geographical variations in market share were recognized but typically simplified through the assumption of a homogeneous distribution of population in space. For applied analysis, this assumption is inadequate in a number of respects. First, population densities are not evenly distributed in space but highly concentrated (most heavily in urban areas). Secondly, population types vary from place to place. For example, in most typical developed cities, the less affluent population sub-groups are concentrated in the centre, the more affluent in the suburbs. A third element which classical analysis fails to address is the idea of spatial disaggregation, where geographies are typically represented by regular tessellations of squares or hexagons. In practice, to build up a picture of small-area variations in the type and density of population, we will be dependent on a variety of administrative geographies for which these data are collected.

In the UK and elsewhere, the primary source of appropriate data is the decennial Census of Population, in which an extensive range of social, demographic, and economic data is collected for all households. Nevertheless, for ease of presentation and the protection of the confidentiality of individual respondents, most of the information is presented by enumeration district (ED), an administrative unit of 50 households or so. From these EDs, we can build to a variety of Census geographies, such as local government wards, parliamentary constituencies, health authority districts, and so on.

A problem with these Census data sources is that government geographies are entirely incompatible with the geographies most relevant to all retail and business clients, which are based on the post-code system. Thus a whole industry has been built up on the conversion of data from its original Census base to business-relevant post-code geographies (for instance, Martin 1991, Raper et al 1992).

A similar split exists in the US between zip-code geographies and census tracts. Other countries have standardized more successfully, either on post-code geographies (for example, the Netherlands) or on administrative geographies (for example, the communes of Belgium or Italy). One feature which all of the geographies share is that they are hierarchical: data are available at very fine spatial scales (such as EDs or 6-digit zips in the Netherlands) but can also be aggregated to a variety of coarser scales, such as postal districts or sectors. For many types of application, disaggregation into districts or perhaps sectors will be more than adequate, but for applications where geographical markets are extremely localized, analysis at something like the ED level may be required. Supermarkets, post offices, and newsagents are examples where this might be true.

Once an appropriate geography for which suitable population statistics are available has been established, then there is a need to establish a method for converting these populations into a demand for goods or services. Four approaches are considered here, based on geodemographics, direct estimation, lifestyles, and direct data.

5.2.1 Geodemographics

The definition of geodemographics was outlined in Section 4.3.5. It was argued there that the major geodemographic packages, such as ACORN and MOSAIC, are used to provide simple descriptive labels for each ED in the UK. Competitive systems have existed in the UK for some time, but the MOSAIC system has only recently been extended into Europe (CCN 1993, Heijt 1994, Birkin 1995). Geodemographics has conventionally been applied to retailing by establishing a link to large market research surveys, such as the British Market Research Bureau's (BMRB) Target Group Index (TGI). TGI is based on a panel of about 25 000 consumers who record ownership, purchase, and preferences for a wide variety of retail and service goods, from fridges to holidays. The address of the panel respondents is used to attach a geodemographic label to each one. It is then possible to produce consumption patterns for each of the different geodemographic neighbourhood types (see Brown 1991 for more details). The geodemographic descriptions of each neighbourhood can be used to map out the variations. An example of this procedure is shown as Figure 5.1. The PiN system has been used to locate customers within drive time bands around Meadowhall – a large, new retail centre which opened close to the city of Sheffield in the UK in 1990. The analyst can estimate the demand for products in these drive time bands.

The strength of these approaches is that they almost invariably present plausible patterns of demand variation between different geodemographic groups. The key problem lies in the linkage process, whereby individual respondents are ascribed a geodemographic characteristic on the basis of the area where they live rather than of the type of person that they are. For example, a BMRB respondent living in an area of 'young married suburbia' might easily be a 65-year-old spinster! This problem, known as the 'ecological fallacy' (see Openshaw 1989b, Flowerdew 1991b, Birkin 1993), is endemic to geodemographics. Whether variations introduced by the ecological fallacy are evened out in use remains an open question.

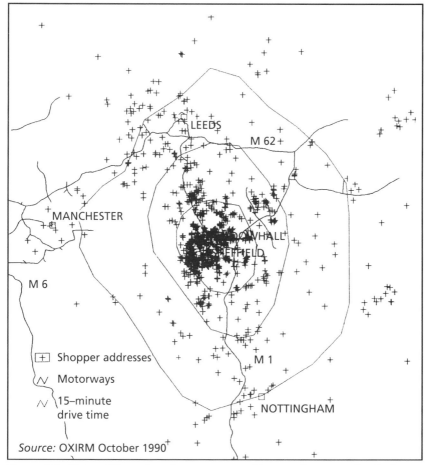

Figure 5.1 Using GIS to pinpoint customer addresses from Meadowhall, Sheffield
(*Source*: Reynolds 1991)

5.2.2 Direct estimation

A second option is to apply market research data directly to small-area population data to derive demand estimates. Many companies collect data on the social class and age profiles of their customers, either through market research or through the capture of point-of-sale data. Census data will easily yield similar population profiles for small areas. The two sets of data can be combined to yield demand estimates. An obvious advantage of this type of approach is its directness – there are no problems with the ecological fallacy, as there are with geodemographics. For this reason it is the main method of demand estimation used in GMAP models.

For example, let us return to the array O_i^{km}. This is disaggregated by region, person type, and product. Imagine that the product we are interested in is records. In the UK, market research data tell us that expenditure on records is highest amongst young people and the higher social grades. From the Census it is possible to extract population by age and social class for each i region in the study (where i could be a Census enumeration ward or a district or a postal geographical region). Then, the appropriate expenditure weightings can be applied to each of these population groups. In the West Midlands, for example, we find the highest concentration of these groups in an area to the north of Birmingham. Hence we expect a concentration of expenditure in this area (see Figure 5.2).

A disadvantage is that to produce estimates, population data must be matched with sector-specific market research data, so general packages cannot be developed to address a variety of products by linking only two databases, as with TGI and ACORN.

5.2.3 Lifestyles

A third source of data with potentially widespread applications in retail marketing is the lifestyle databases, launched in the mid-1980s. Questionnaires of various types are used to record consumer behaviour, and the responses are entered into a database. As the databases have become larger and larger, they can be used simply to profile the populations of small geographical areas. The largest UK lifestyle database, as mentioned in Chapter 4, is NDL's Lifestyle Selector, with information on over 10 million households: roughly a 50 per-cent sample of the whole population. While rather biased in its composition towards the most active consumer groups, responses may be 'reweighted' using small-area census data (Birkin 1993, Birkin and Clarke 1995) to provide the best of both worlds.

There are at least three different ways in which this type of data can be valuable. First, the lifestyle questions go beyond the traditional Census-type questions regarding occupation, demographics, and so on, to ask about behaviours, hobbies, and preferences. Thus we can get a direct picture of things like spatial variations in book-reading, or bingo-playing, which could clearly be used in trying to establish potential markets for these products. Secondly, a particular variable which is not collected elsewhere is household income, which is almost certain to be key in describing small-area expenditure patterns for any product. Indeed, Webber (1992), who was responsible for the initial development of ACORN and MOSAIC, has tentatively shown that the income variable within lifestyle databases may be a more powerful discriminator than ACORN or MOSAIC. Thirdly, lifestyle databases may be used to focus on the 'aspirational' component to purchasing decisions. Psychographic classifications, such as CCN's

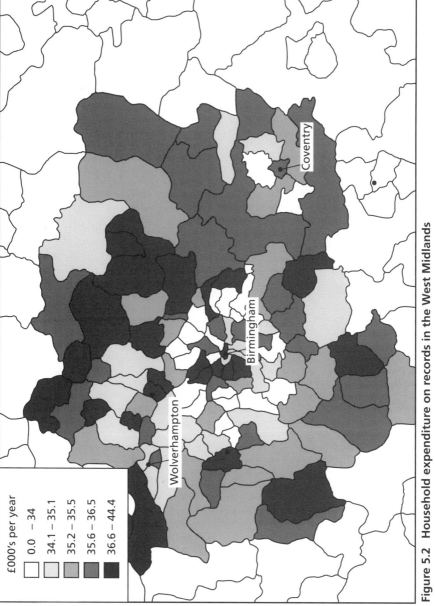

Figure 5.2 Household expenditure on records in the West Midlands

Persona, focus on the idea that certain types of individual may be more receptive to new products than others. These principles may also be applicable to cross-selling: someone interested in hang-gliding might be more likely to purchase an XR2 than a Ford Fiesta, for instance.

5.2.4 Direct data

In some retail markets there is no need to estimate demand by small geographical region: the data already exist. This is very rare, however, and normally occurs only when the ownership of a product has to be registered with the appropriate authorities. A good illustration (and one we shall discuss in detail in Chapter 9) is the demand for new cars. In almost all countries of the world, a new car purchase has to be registered for tax and licence purposes. In many instances these data are computerized, and usually sold back to the motor industry. For fine geographical regions (usually postal zones or communes), manufacturers can therefore obtain the number of new cars registered in that region for both their own organization and the competition. If time series data are available, then this forms a much more reliable picture of the demand for new cars by small geographical region than any method to estimate that demand by linking population characteristics to likely buy rates associated with different age groups, social class groups, etc.

5.3 THE SUPPLY SIDE

In order to build effective models of its markets, a retailer needs to understand who and where its competitors are, as argued in Chapter 4. In the models described in Section 5.1 there are two variables to be defined: W_j^m and θ_j^{mn}. The first term was described as the attractiveness of shopping centre j for products type m. Thus for each application we need to identify the nature of the market (all those outlets selling product m) and their locations. Second, we need to measure the attractiveness of these outlets. We shall look at each of these in turn.

Assuming that an organization is able to identify its competitors, the problem of locating them can vary enormously from sector to sector. For example, in the car market, each manufacturer typically maintains an up-to-date dealer list for marketing purposes, so a good database of dealer locations can be assembled. Because financial service outlets share a common service network (so that it is possible for a customer of one bank to deposit a cheque with a branch of a completely different bank), various 'sort-code directories' exist to allow the efficient transfer of funds to take place. In the US, individual states maintain comprehensive restaurant inventories for the purposes of taxation and public health monitoring.

In the retail sector at large, comprehensive and accurate information is typically more difficult to obtain. Although attempts have been made to produce directories in the UK (for example, by Newmans and Chas E. Goad), the retail sector is so diffuse and volatile that no one directory source may be viewed as accurate and up to date. For example, Goad's shopping-centre database covers only around 1000 centres, and the smaller ones are updated only once every two or three years. At a conservative estimate there are probably twice this number of centres of retail activity in the UK, and peri-urban (edge-of-town) activities are difficult to monitor. These directories therefore need to be supplemented by sources such as telephone directories (in computerized form, of course!) and, where possible, in-house market intelligence. For example, in the UK the *Yellow Pages* database lists entries by category type (for instance, do-it-yourself or DIY outlets) along with post-code locations.

Once the competitor network has been identified, is it time to address more specific questions about individual units. What is their precise location? Databases will typically exist (such as the Postcode Address File in the UK) which allow addresses to be geocoded, but problems may arise, especially if competitor inventories are pulled together from a variety of sources. How does one tell, for example, whether two stores with a similar address in different databases are the same or distinct? This can be a particular problem in the US, where one might easily find several stores of a single chain on Main Street in a large city. A related issue is how to determine what comprises 'shopping centre j'. In the grocery market, it is possible to model the flows between demand zones and all major outlets (supermarkets and hypermarkets) explicitly. In this case, shopping centre j should really be defined as store j. However, when the number of potential supply points increases it becomes necessary to think about aggregating individual outlets into the traditional concept of a shopping centre or into retail malls or parks. This is most likely to be the case when considering comparison goods.

In addition, there are a number of activities which are traditionally located outside shopping centres. For example, there are estimated to be 45 000 CTNs (confectioners, tobacconists, newsagents) in the UK alone, and over 11 000 car dealers. So there is a problem of whether to group these competitors in some way for analytical purposes, and if so, how. For mainstream retail activities, it may be appropriate to group outlets into continuous shopping centres, but this is unlikely to be appropriate for car dealers. On the other hand, even car dealers cluster together, and such a dealer cluster may have a meaningfully distinct identity. An example where both types are likely to combine is in out-of-town retailing, such as DIY stores in the UK. Here large outlets may be located on stand-alone sites, typically on the edge of town. Alternatively, several large competitors may be clustered together on a retail park. Finally, they may compete with other retailers in conventional high street locations. This situation is illustrated in Figure 5.3. Each application has to consider the definition of shopping centre j very carefully.

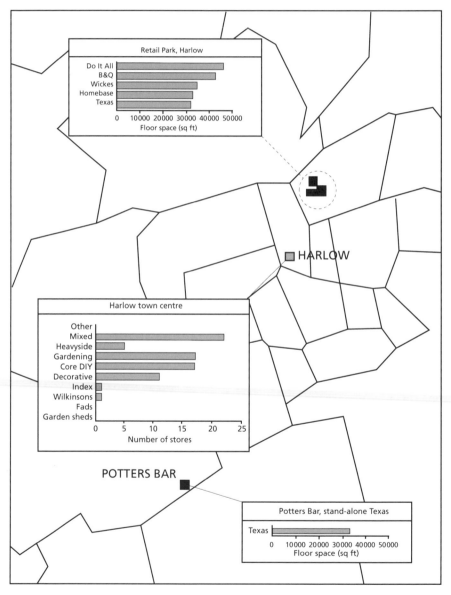

Figure 5.3 Variety of retail centres for DIY products in west Essex and north-east London

The second consideration in the array W_j^m is the measure of store attractiveness. In standard applications, floor space is the most common proxy for attractiveness. Indeed, a number of studies have found floor space the most *statistically* reliable proxy for attractiveness (Pacione 1974). Store sizes are not as readily available as one might think. Although map-based directories such as Goad may give an indication of the size of competitor units, detailed field research is usually the only way to make progress here. Typically this is not a problem in the supermarket business, where the major players are preoccupied, perhaps to the point of obsession, with monitoring the size and location of new and existing competitor outlets. In other areas, especially where franchising is involved (for example, in the car market), it may be difficult for a retailer to assess the size of its 'own' outlets, let alone those of its competitors. At the very least, however, it is usually possible to obtain average company store sizes from gross statistics revealed in market reports or the company's own annual reports.

The second variable we may wish to measure in the model is θ_j^{mn}. Remember that this variable has been defined as the attractiveness of retailer n (over and above differences in floor space). This variable attempts to capture something about company performance, image, and pricing. For example, a Suzuki car dealer in the UK does not sell the same number of new cars on average per dealer as a company such as Ford or Rover. However, the estimation of competitor performance is a very difficult task. Again, it is often difficult enough for a retailer to keep track of the sales of its own branches, let alone those of its competitors; but from market reports, it is possible to say something about retail performance given the number of outlets. Table 5.1, for example, shows the level of revenues per square foot of grocery trading floor space in the UK in 1992. These sorts of difference in average outlet performance can then be directly input into θ_j^{mn} to produce variations in attractiveness by retailer.

5.4 SPATIAL INTERACTION

There are typically two classes of interaction data required to operationalize the models described in Section 5.1. The first is customer data on actual levels of interaction. This is required in order to test whether our models are reproducing existing customer behaviour satisfactorily (a process called calibration – see Section 5.5). Typically the institutions with the best interaction data will tend to be those such as banks or rental organizations (for example, TV and video rentals) which have account-based activities and regularly capture customer data. Individual branches know the locations of their customers because they need to send bills or accounts on a regular basis. Notice that in this situation much less will be known about the behaviour of competing organizations, so the interaction patterns still have to be generalized in some way.

Table 5.1 Sales intensities of multiple grocers

Fascia	Sales intensities (£ per sq.ft per week)
Sainsbury	17.46
Waitrose	16.89
Tesco	13.29
Morrison	11.60
Safeway	11.38
Somerfield	11.20
Wm Low	9.36
Asda	9.24
Netto	9.00
Kwik Save	8.90
Lo-Cost	8.68
Budgen	8.64
Presto	8.58
Gateway	7.90
Iceland	7.31
Food Giant	4.50

Retailers who lack comprehensive information will probably have to get by with some form of survey sampling. This might involve stratified interviews in the home (door-by-door or by telephone); in-store interviewing; or some sort of shopping-centre survey outside of the store. Often these procedures may be reinforced through special promotions, where customer demographics are collected in tandem with a free draw or special offer. There is a place for each of these techniques in putting together a comprehensive picture. In this case, we need to generalize from a sample of a particular retailer's customers to the whole universe; although we may also take the opportunity to generate information about competitor interactions through the survey process.

The possibility of acquiring interaction data automatically, for example through the use of smart cards (where, say, the customer pays for groceries with a credit card which also holds his or her address: the computer strips the address as the payment is debited), comes ever nearer. Pilot schemes have now been undertaken in the US, Japan, and many European countries.

Another important class of interaction information concerns measures of spatial impedance. Our objective is to build a model which represents customer patronage of specific stores, which we wish to relate to the availability of retail opportunities. To make progress, we need to gauge the ease of travel to various opportunities. That is, we need to estimate the array $\{c_{ij}\}$. At one extreme, it is straightforward to identify a straight-line distance between customers and retailers (subject to certain aggregation difficulties, such as how to represent the 'average' location of a customer within a postal sector), but this is unlikely to represent impedance realistically in many situations (for instance, if the two are separated by a river which cannot be crossed). At the other extreme, generalized trip costs could be considered: how long does it take? How much petrol is needed? What is the car-parking charge? All of these could then be converted to monetary equivalents (if necessary) and added together. This is feasible in some circumstances, and has been common practice in many transport modelling studies, but is probably an unnecessarily involved process for retail modelling purposes. A sensible intermediate option is to consider the drive time between origins and destinations. Appropriate sets of drive times can be generated within most major GIS packages from road network data, subject to appropriate assumptions about traffic speeds (or other impedances) along the network links.

5.5   MODEL CALIBRATION

The key component of the modelling process to which we now turn is calibration: the process by which numerical values are assigned to the model parameters in such a way that the model accurately reproduces the real patterns. Typically we need to consider two components of the calibration process: β calibration and revenue calibration.

β calibration involves key parameters which govern the general 'pattern' of the interactions. High values of β are associated with short travelling times or distances; low values of β give patterns which are much more dispersed. From the model equations, β values might be expected to vary by origin zone, by person type, and by product type. For example, one would expect to find lower average trip lengths and more concentrated patterns for low-value products than high-value ones, for obvious reasons: compare Figures 5.4a and 5.4b for the difference in catchment area by product. Typically, more affluent social groups might be expected to be less sensitive to distance (as they are more likely to have cars at their disposal, and so on), and country dwellers to be more likely to travel long distances than city residents.

The methods for calibrating the distance deterrence parameters are well documented (see, for example, Williams and Fotheringham 1984). In general, the objective is to minimize the difference between observed and predicted flow patterns, although usually this is best achieved after some transformation to reweight the flows, such as the maximum likelihood method:

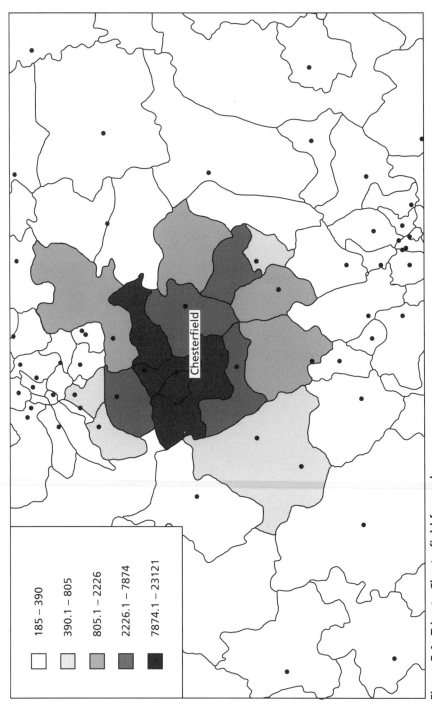

Figure 5.4a Trips to Chesterfield for cards

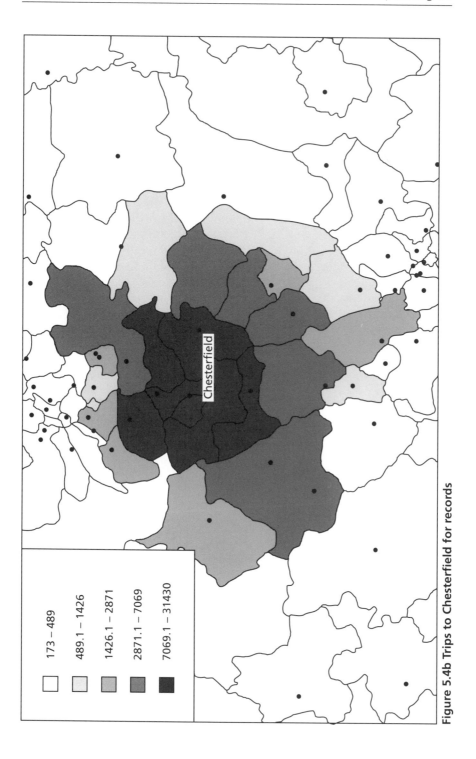

Figure 5.4b Trips to Chesterfield for records

Chesterfield

173 – 489

489.1 – 1426

1426.1 – 2871

2871.1 – 7069

7069.1 – 31430

$$Min_{\{\beta_i^{km}\}} \sum_j \frac{\log S_{ij}^{km^*} - \log S_{ij}^{\#km^*}}{\log S_{ij}^{\#km^*}} \tag{5.3}$$

where $S_{ij}^{\#km^*}$ represents the *observed* interaction flows. Note that, as we saw in Section 5.3, the observed interaction data are unlikely to be comprehensive, and so these data must be carefully reweighted to remove potential biases before any calibration is attempted.

The maximum likelihood estimation process allows us to find the best set of parameters to match to observed data given an appropriate model specification, such as (5.1) and (5.2). It is also important to make a judgement of the adequacy of the resulting patterns. How similar are the observed interaction patterns to the model predictions? Examples are shown in Figure 5.5. Clearly the patterns in part (c) match rather well; those of part (d) are much less satisfactory. One of the problems is how to quantify this goodness of pattern fit, and how to identify any anomalies, without mapping out patterns for all possible origins and destinations. Quite often, strange patterns might be caused by data errors, such as rogue drive times, which can be corrected easily once trapped.

A whole battery of goodness-of-fit statistics for spatial interaction models has been proposed and tested in the literature. These include regression statistics (for example, the correlation coefficient of observed against predicted interactions), information statistics (such as minimum discriminant information, entropy statistics, and standardized root mean square error). A comprehensive review of the possibilities is provided by Knudsen and Fotheringham (1986). One of the problems with all of these statistics is that significance testing is very difficult. Put another way, this means it is difficult to tell from the statistics whether a predicted interaction pattern looks 'like' the observed pattern. Fotheringham (1992) is one observer who has already noted some potential uses of GIS in pattern matching exercises of this type. More generally, there is clearly scope for the implementation of artificially intelligent pattern recognition techniques here.

Two simple but useful statistics which can be used here are Spearman's rank correlation coefficient and the Wilcoxon matched pairs test. These statistics can be produced as a measure of pattern fit for each individual residential area or destination. Spearman's technique involves ranking potential destinations from most to least likely on both model and observed data and producing a measure of the correlation between the two lists. The Wilcoxon test looks at the distribution of flow sizes between the two interaction sets. For example, if the observed data have two or three large flows and a small number of small flows, while the predicted interaction matrix has many medium-sized flows, then a high Z-score results. In the example patterns shown in Figure 5.5, the dealer in Lewisham (Figure 5.5c) scores

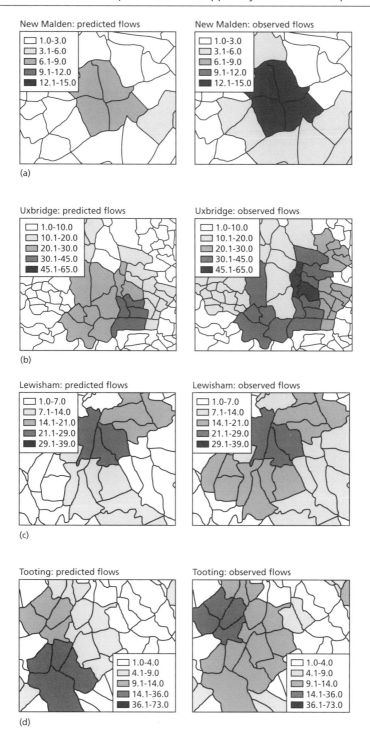

(a)

(b)

(c)

(d)

Figure 5.5 Observed versus predicted flows to retail centres

well on both the Spearman's and Wilcoxon tests. On the other hand, the dealer in Tooting (Figure 5.5d) fares poorly on both counts – there are clear problems with the model here. The dealer in New Malden (Figure 5.5a) has a good Spearman's rank but a poor Wilcoxon, showing that trade is being pulled from the right market area but in an unbalanced way. The accuracy of the modelled drive times to this dealer might need to be investigated. Finally, the dealer in Uxbridge (Figure 5.5b) has a good Wilcoxon score but poor Spearman's rank, showing well-balanced flows, but not from the right postal sectors. Often this type of situation can arise, where a main dealer has a strong sub-dealer who is selling a lot of cars in his or her own locality, but they are being registered through the main dealer.

For revenue calibration, the spatial interaction models are constrained to a set of small-area expenditure estimates. The models are not constrained in the same way at the destination end. The interactions can therefore be summed to provide revenue estimates for the retailers:

$$D_j^{mn} = \sum_{ik} S_{ij}^{kmn} \tag{5.4}$$

where D_j^{mn} is the revenue of retailer n for product m at centre j.

Since one of the uses of the model is likely to be in sales forecasting (see below, Section 5.7), it is clearly desirable also to calibrate these revenue estimates to known data. Usually it will be appropriate to focus on something like total relative error (TRE):

$$TRE = \sum_{jm} \frac{(D_j^{m1} - D_j^{\#m1})}{D_j^{\#m1}} \tag{5.5}$$

where $n = 1$ is the client organization.

The revenue estimation process is itself clearly tied to market shares within a centre, so it will usually be appropriate to develop further models which relate variations in market share to considerations such as outlet size and pitch, layout, local brand strength, local demographics, and so on. Appropriate relationships can be established through any of a variety of statistical techniques, such as multivariate regression, discriminant analysis, or even neural networks.

An example of this process, drawn from a real application, is shown in Table 5.2. The names and numbers have been disguised to maintain client confidentiality. Nevertheless, the results have not been exaggerated or distorted significantly. GMAP was invited to specify and calibrate a model for a small region of Britain containing roughly 2 million households and 14 client stores. As a test of model accuracy, revenue data were provided for eight stores and withheld for the remaining six (shown in bold type). As the table shows, GMAP was able to estimate the missing revenues to an accuracy of ±7 per cent. Although it is not untypical for clients to wish to

Table 5.2 Model results for 14 stores in south-east England

Centre name	Sales ratio	Model prediction	Errors (%)
Ashgate	200	220	10.0
Bayfield	160	150	6.3
Chetburn	120	118	1.7
Darkridge	105	125	19.0
Eastport	100	102	2.0
Fernley	90	83	7.8
Greenthorpe	80	75	6.2
Holford	60	50	16.6
Icklington	**185**	**180**	**3.0**
Jaywich	**147**	**150**	**2.0**
Kentside	**105**	**120**	**12.0**
Ladyhill	**105**	**100**	**5.0**
Millwood	**84**	**85**	**1.0**
Northdale	**72**	**70**	**3.0**

withhold store revenues in this way as a test of model accuracy, this example represents both the most stringent test, and also the most successful outcome, of such an exercise. In general, our experience is that to obtain turnover estimates within ±15 per cent about 85 per cent of the time is a reasonable target for mainstream retail sectors. This experience appears to be shared by some other modellers attempting to reproduce store sales (such as Janssen 1994), for reasons which will be discussed in Section 5.6. Notably better results have, however, been claimed for a gravity-type model of supermarket sales developed in-house by Tesco (Penny and Broom 1988).

In our experience, other sectors are typically even less predictable, with the least predictable of all being car dealerships. A well-performing dealer, it appears, is able to sell cars in all but the poorest of locations, while poorly performing dealers are frequently unable to exploit fully even the most gilt-edged of opportunities.

Returning to Table 5.2, the argument is that if it is possible to reproduce observed outlet sales to such a degree of accuracy, and in a manifestly unbiased way, then we can expect the same level of model performance in forecasting new store potentials. The benefits of this level of forecasting are obvious in terms of both network optimization and new store development opportunities. This is illustrated schematically in Figure 5.6. Suppose that in order to open a new store, 'hurdle' sales of £800 000 are required to break even. If an existing technique has an accuracy of, say, ±30 per cent, then even a new opportunity with forecast sales of £1 000 000 cannot be guaranteed to clear that hurdle.

5.6 ANALYSIS: PERFORMANCE INDICATORS

One potential application of the baseline model can be seen if we return to the discussion of revenue estimation which we left in Section 5.5. When a model has been well calibrated, it may be possible to interpret variations between observed revenues and the model predictions as indicators of *store performance*. That is, there are clearly some factors which one would not usually attempt to represent within the models, such as the strength of management within a store.

Another simple example can be seen by returning to Table 5.2. One of the stores for which the model is sizeably over-predicting the turnover potential is Darkridge. It later transpired that actual sales for the store in question were not helped by major road repairs blocking access to the store for a period of at least nine months! Of course, it is futile to try and reproduce this sort of behaviour within the models themselves, and there are many reasons why individual stores might exceed or fall short of their potential. Any comparison of modelled turnover potential against actual sales can almost always be expected to shed interesting light on the performance of the store network.

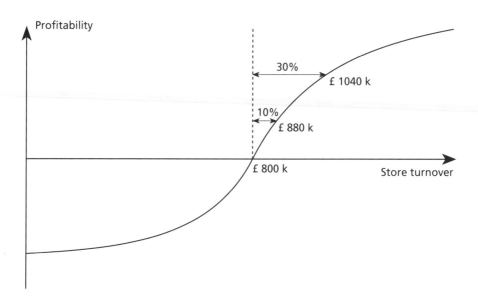

Figure 5.6 Importance of accurate revenue projections for new store development

Considerable progress has been made recently with performance indicators from spatial interaction modelling through the definition of suites of residence-based and facility-based indicators (see Bertuglia et al 1994a). Residence-based indicators relate to how well residents of a locality are served. A key residential performance indicator is market penetration (MP_i^{mn}), defined as an individual retailer's share of the market by small geographic area (such as a postal sector):

$$MP_i^{mn} = \sum_{jk} S_{ij}^{kmn} \Big/ \sum_{jkn} S_{ij}^{kmn} \tag{5.6}$$

As we have already seen, most organizations can be expected to have a good idea of their turnover at different outlet locations; and it is not difficult to estimate demand by small areas with a certain degree of confidence. However, only by estimating the flows from individual residential areas to retail centres can we tie the two elements together, to estimate market penetration by residential area. The process usually requires the calibration of a spatial interaction model. In practical applications, analysis of market penetrations will usually throw up interesting insights at a variety of spatial scales. At a regional level, it is well known that supermarket retailers such as Sainsbury's and Tesco are strongest in the south of England, while others, such as Wm Morrison, Kwik Save, and Asda, are stronger in other regions. The recent battle for the Scottish retailer Wm Low between Tesco and Sainsbury's is an obvious strategic response to this imbalance. Equally important, however, but less widely recognized are small-area variations in market penetration. While this is again a well-known phenomenon for supermarket retailers, it applies equally to high street retail, car dealers, banks, and building societies. Examples are given in Chapters 9 and 10.

The concept of market penetration is a useful one, not least because it brings together demand and supply. This is not easy to do, because supply-side characteristics relate to a discrete set of centres which sit on a continuous but essentially unrelated set of residential areas. Another way in which modelled interactions can be used to produce a residence-based indicator which balances demand and supply is through the concept of *effective delivery*. The indicator would typically be defined as follows (noting here that we have dropped the superscripts for ease of understanding, although in reality the indicator is that much richer for being disaggregated by person type, product, etc.):

$$ED_i = \sum_j Z_j \frac{S_{ij}}{\sum_i S_{ij}} \tag{5.7}$$

where Z_j is a measure of the level of retail activity at j, typically floor space. The indicator is constructed by sharing the floor space at each centre between the residential areas in the catchment of the centre. The result is a measure of the retail floor space available or 'delivered' to the residents of that area. If this level of delivery is normalized according to the size of the residential populations we obtain the 'provision ratio':

$$PR_i = \frac{ED_i}{P_i} \tag{5.8}$$

In public sector modelling applications, such as health-care delivery (see also Chapter 6), this measure comprises a useful indicator of equity, that is, how fairly facilities are distributed between geographical areas. While equity is typically not a priority for providers of private sector services, such as retailing, nevertheless variations in provision ratio may be a useful pointer to potential development opportunities. Consider, for example, Figure 5.7 (Plate 5.1), which shows the level of provision of supermarket floor space across postal sectors in Yorkshire. While most of the major markets are heavily and fairly evenly penetrated, the Hull area, particularly to the west, has noticeably lower levels of provision. A good working hypothesis would be that there are profitable opportunities for further development in this area. Of course one would want to test this hypothesis further by specific 'what if?' analyses in the manner of Section 5.7.

If the provision ratio is seen as a measure of equity across demand areas, then it is also possible to build up similar measures across centres (that is, facility-based indicators), largely through the concept of 'catchment populations'. The catchment population is a measure of the number of shoppers using each centre, and can be calculated as:

$$CP_j = \sum_i P_i \frac{S_{ij}}{\sum_j S_{ij}} \tag{5.9}$$

One can then derive measures such as floor space or aggregate centre sales per head of catchment population, again as an indicator of imbalances between supply and demand. When all the superscripts are added, this will show different catchment areas for different retailers, product types, etc.

Other useful indicators can be generated using catchment populations as an averaging mechanism, such as average expenditure per head of catchment population:

$$a_j = \sum_i e_i P_i \frac{S_{ij}}{\sum_j S_{ij}} \tag{5.10}$$

This represents a convenient way of representing residential characteristics at the centre level. This type of indicator is particularly important if we are to try to develop secondary models of store performance. For example, one would typically expect to find a retailer like W.H. Smith (WHS) in centres with a higher average expenditure per head than, say John Menzies, because market research shows WHS draws more heavily than Menzies from the 'higher' social classes. When trying to interpret the performance of WHS stores, do we find that the stores typically do better in the more affluent centres? If so, then to expand into centres with less affluent populations would probably be a poor strategy. The reverse would presumably apply to Menzies.

5.7 FORECASTING

The most obvious type of scenario planning capability offered by the retail decision support systems described in this chapter is impact analysis for new store openings. It has been shown above that appropriate spatial interaction models can be calibrated to represent retail interaction patterns both accurately and robustly. In this situation, the models can be extrapolated to allow for marginal changes in the retail environment. The impact of any of the variables which influence the models can be considered in this way: for example, the effects of a new road or bus route on transport costs might affect the patterns; or a new housing estate could affect local demand. However, the most important effect to model is the likely impact of a new store. This impact analysis can have a number of different dimensions. The primary concern will be to establish the potential revenue-generating capabilities of a new store, and to compare this to the investment required to determine the net benefit from a new development. It is important to remember, however, that new stores can also have a negative impact on the existing store network by cannibalizing trade. Often, this cannibalization can make the difference between viability and non-viability for a new proposal.

Consider the example shown in Table 5.3. Again, this example is hypothetical, but based on a real application of the models. Five new locations for new quick-service restaurants are being considered. Although the rents associated with the different sites will vary somewhat, a sales level around $800 000 per year can be considered as a reasonable target for viability. The model immediately shows that Stores 1 and 2 may be non-viable on revenue terms alone. Of the others, Store 3 has good revenue potential but impacts so seriously on nearby outlets that the net effect to the organization is not beneficial. Only Stores 4 and 5 are fully viable: Store 5 contributes most in terms of bottom-line profit, even though its revenue only ranks third of the five stores considered.

The models can also be used to show where customers are likely to come from to the new store, and to show the effect of a new store opening on

Table 5.3 Impact analysis for five new quick-service restaurants (*Source*: GMAP 1994)

	Store 1	Store 2	Store 3	Store 4	Store 5
Total revenue	733	790	916	1004	908
Major impact	49	62	83	82	22
Total impact	54	82	127	132	32
Net revenue gain	679	708	789	872	876

Notes: *Total revenue* is the expected revenue for Stores 1 to 5 (in $000s per year)
Major impact is the amount of revenue it is expected will be lost at the outlet which is worst affected by the new store opening.
Total impact is the amount of revenue which it is expected will be lost at all other outlets in the network.
Net revenue gain is the total revenue less the total impact.

local market penetrations. Consider the example shown in Figure 5.8, where the town of Hastings on the south coast of England has been chosen for a new store by the retail group Dino-Stores. The baseline map shows that the group is able to achieve penetrations of anything between 20 and 30 per cent in small areas on the coast. There are large tranches of land with much lower penetrations. The scenario map shows that these gaps are only partially filled by the new store development. The argument here would be that there are probably many other, and possibly better, opportunities for new stores elsewhere in this region; and also that local marketing promotions for the new store, such as door-to-door coupon drops, billboard advertising, or use of local media, need to be strongly targeted spatially.

Before we leave the subject of new store impacts, it is important to note that the technology can also be used to gauge the potential effects of new store openings by competitors. This is particularly important for retailers such as supermarkets, which need to consider that when a particular site becomes available then if they do not get it, one of their competitors very probably will. The same could easily apply to retailers that are offered space for rent within new malls or arcades. All of this type of planning, while important, can be thought of as reactive: it represents a process by which retailers are offered a new opportunity by property developers and the like, and must decide whether to exploit it. An important feature of our systems is that they allow potential opportunities to be evaluated much more rapidly and accurately than was previously possible. However, another important feature is that they can also facilitate a much more proactive approach to the problem. By studying maps of existing market penetrations, gaps in

Baseline penetrations

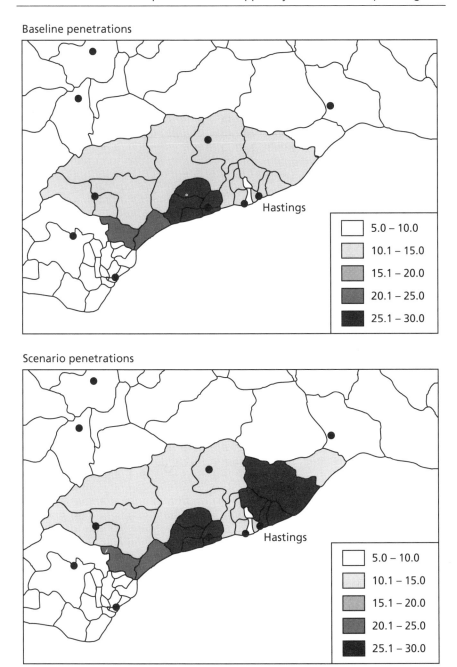

Scenario penetrations

Figure 5.8 Change in market penetrations for a new Hastings store
(*Source*: GMAP 1994)

the existing retail network can be identified. If something like provision ratios are overlaid, areas where the competition is weak can be identified. In the midst of all this it may be possible to begin to search for new opportunities, and only then go about the process of finding appropriate sites: a much more efficient and effective approach to the problem.

As we have seen above, one of the key drivers of the spatial modelling system is store attractiveness, which will usually be strongly related to floor space. One of the ways in which this property can be exploited in predictive analyses is through *space planning*. This is a common name for the process by which retail managers try to optimize the product mix within stores to find the best balance between lines. Consider Bassetts stores as an example. The company sells a diverse range of products, from clothing and footwear to cosmetics and jewellery, and operates different fascias including Harvey's (a wide range of products, but especially clothing) and Icebox (jewellery). So if this organization is considering redevelopment of an existing site, a number of questions arise: is the current fascia the most appropriate? What is the ideal product mix to offer within that store? Which changes have most impact on the other stores in the chain?

An example is shown in Table 5.4, where the company is considering reformatting its existing Harvey's store in the small market town of Castlebridge. The existing store has a good balance between the different products, and a gross annual turnover of just under one million pounds. There is also an Icebox store in the centre with a turnover around £400 000 for jewellery. The Harvey's store has to be moved into a new and smaller unit: the question is how to manage the reduction in space? In the first scenario (Table 5.4a), the loss of space has been split between children's clothing, haberdashery, and cosmetics, with the biggest reductions in the last category. In this case there is, unsurprisingly, no impact on the Icebox store. In the second scenario (Table 5.4b), the jewellery section of Harvey's has been eliminated completely. While the allocation for cosmetic products has also been reduced (less dramatically than previously), the space for children's clothing and haberdashery has now been *increased*. The key outcome of this new scenario is that the removal of jewellery allows some of the available jewellery expenditure in Castlebridge to be picked up by Icebox – in effect, a 'negative deflection' process – and this makes the second scenario more attractive in net revenue terms for the Bassetts Group as a whole.

In the late 1960s, when mathematical models of retail systems first began to gain popularity in the UK, one of the most prominent uses of the technology was in the evaluation of the impacts of new out-of-town shopping centre proposals such as the Haydock scheme (Foot 1981). Such applications remain relevant today as a new wave of out-of-town developments is

Table 5.4 Effects of a store reformat in Castlebridge (*Source*: GMAP 1995) (a) Castlebridge 1

Store: Harvey's

	Gents' clothing	Ladies' clothing	Children's clothing	Jewellery	Haberdashery	Cosmetics	All
Base revenue	199.0	216.4	241.8	78.7	86.0	88.0	909.9
Base floor space	40	144	74	34	35	32	359
Base market share	16.6	14.8	74.6	11.6	18.8	36.4	20.9
Model revenue	199.0	216.4	228.7	78.7	76.9	49.9	849.6
Model floor space	40	144	68	34	31	17	334
Model market share	16.6	14.8	72.9	11.6	17.0	23.3	19.7

Store: Icebox

	Gents' clothing	Ladies' clothing	Children's clothing	Jewellery	Haberdashery	Cosmetics	All
Base revenue				417.6		26.0	443.6
Base floor space				78		7	85
Base market share				61.7		10.7	48.3
Model revenue				417.6		27.8	445.4
Model floor space				78		7	85
Model market share				61.7		12.9	50.0

Table 5.4 (b) Castlebridge 2

Store: Harvey's

	Gents' clothing	Ladies' clothing	Children's clothing	Jewellery	Haberdashery	Cosmetics	All
Base revenue	199.0	216.4	241.8	78.7	86.0	88.0	909.9
Base floor space	40	144	74	34	35	32	359
Base market share	16.6	14.8	74.6	11.6	18.8	36.4	20.9
Model revenue	199.0	216.4	269.5	0	99.3	65.6	849.8
Model floor space	40	144	88	0	41	23	336
Model market share	16.6	14.8	77.7	0	21.3	29.1	19.5

Store: Icebox

	Gents' clothing	Ladies' clothing	Children's clothing	Jewellery	Haberdashery	Cosmetics	All
Base revenue				417.6		26.0	443.6
Base floor space				78		7	85
Base market share				61.7		10.7	48.3
Model revenue				440.2		27.0	467.2
Model floor space				78		7	85
Model market share				69.8		12.0	54.6

Table 5.5 Bluewater Park: centre turnover by product (*Source*: GMAP 1994)

Centre: Bluewater Park	Gents' clothing	Ladies' clothing	Children's clothing	Jewellery	Haberdashery	Cosmetics
Centre floor space	307	2623	914	1358	1160	570
Centre revenue	1724	4217	848	3748	1273	848
Bassetts floor space	127	517	294	151	140	103
Bassetts revenue	801	924	464	320	247	221
Bassetts market share	46	22	55	9	19	26
Icebox floor space				192		22
Icebox revenue				674		42
Icebox market share				18		5
Harvey's floor space		1000				
Harvey's revenue		2077				
Harvey's market share		49				
Greyhound floor space						670
Greyhound revenue						2175
Greyhound market share						47

beginning to take place (although these were slowed to a great extent by the economic recession of the early 1990s). Examples include the Meadowhall centre near Sheffield, Merry Hill in the West Midlands, and the Thurrock and Bluewater Park developments to the east of London. Although mathematical modelling techniques are rarely used for such purposes now (for the kinds of reason which will be explored in Chapter 7), they remain clearly relevant.

Consider the case of Bluewater Park, a development near Swanscombe on the south side of the Dartford Tunnel to the east of London. Table 5.5 shows an attempt to gauge the potential turnover of the centre for the key products offered by Bassetts stores. Like other developments of this type, its size is very large, generating square footage and associated turnover on a par with the larger town centres in the region. This is illustrated in Table 5.6, where we also see that the large towns of Gravesend and Dartford lose huge amounts of revenue, equivalent to between one fifth and one third of their turnover without Bluewater Park. This analysis starts to look uncannily like the results of the Haydock Study, where the proposals were eventually turned down largely because of projected impacts of this magnitude on local town centres.

The problem for Bassetts is that many of the centres most severely impacted upon by the new centre contain their own stores. So the big deflections feed through into significant cannibalization of existing trade, as shown in Table 5.7. Overall, of the £7.9 million in revenue which is generated in this particular scenario, no less than £4.3 million is diverted

Table 5.6 Bluewater Park: centre deflections (*Source*: GMAP 1994)

Centre	Baseline revenue (£000s)	Scenario revenue (£000s)	Difference (£000s)
Gravesend	8 946	6 189	2 757
Dartford	10 296	8 187	2 109
Thurrock	13 871	12 359	1 512
Bexleyheath	12 093	11 188	905
Maidstone	24 877	24 159	718
Swanscombe	1 714	1 134	580
Chatham	11 835	11 264	571
Orpington	9 361	8 961	400
Hartley	1 157	762	395
Rochester	7 709	7 358	351
Strood	2 126	1 918	208
Sevenoaks	4 884	4 709	175
Bromley	18 027	17 854	173
Sidcup	6 702	6 535	167
Hempstead Valley	3 218	3 108	110

from existing stores elsewhere in the region. However, the final subtlety is that much of these existing revenues would be lost even if Bassetts were to stay out of Bluewater Park, as their competitors will still be there. For example, Table 5.5 shows that a new Icebox store in Bluewater Park can be expected to generate £0.7 million in revenue, but Table 5.7 indicates that existing Icebox stores in the surrounding centres will lose a total of £1.5 million. The loss in revenue is much greater than the incremental revenue generated, because new revenue is also being generated by other jewellery stores on Bluewater Park. The model can be run to show that if all Bassetts stores are withdrawn from Bluewater Park, then £2 million in lost trade is taken by the competitors who have opened there. Thus the marginal impact of opening the stores shown in Table 5.7 is in fact £2 million less than the £4.3 million deflected in the primary scenario, that is, 'only' around £2.3 million. Clearly this still represents a huge and potentially crucial sum in assessing whether Bassetts should go ahead and move into the new centre.

Table 5.7 Bluewater Park: store deflections (*Source*: GMAP 1994)

Store	Baseline revenue (£000s)	Scenario revenue (£000s)	Deflection (£000s)
Harveys:			
Bexleyheath	626	567	61
Bromley	1317	1303	14
Maidstone	1161	1100	61
Greyhound:			
Chatham	652	613	39
Dartford	371	303	68
Thurrock	575	521	54
Bassetts:			
Bexleyheath	3453	3179	274
Brentwood	1362	1344	18
Bromley	3880	3846	34
Chatham	2079	1962	117
Icebox Dartford	1375	1023	352
Eltham	1658	1631	27
Gillingham	965	931	34
Gravesend	1582	912	670
Hempstead Valley	861	840	21
Maidstone	3376	3262	114
Orpington	2564	2401	163
Sevenoaks	1677	1594	83
Thurrock	3321	2987	334
Tonbridge	1353	1333	20
Tunbridge Wells	3398	3378	20
Icebox:			
Bexleyheath	904	776	128
Bromley	1252	1219	33
Chatham	1101	990	111
Chelmsford	1048	1036	12
Dartford	568	344	224
Eltham	521	502	19
Gravesend	604	257	347
Hempstead Valley	866	813	53
Maidstone	924	776	58
Thurrock	1743	1443	300
Tonbridge	433	416	17
Tunbridge Wells	1152	1134	18

5.8 CONCLUDING COMMENTS

In drawing together some conclusions from this chapter, it is perhaps appropriate to return explicitly to the subject of GIS. The nub of the argument here must be that the development of systems for retail clients which integrate expenditure data, competitor locations, and interaction information clearly represents a large potential market for GIS. However, in order to exploit this opportunity, GIS needs to do more than offer neat technical solutions to the problems of data management, and attractive computer-generated maps when the data have been synthesized. Rather, the GIS must be used as a platform in which different strategies can be evaluated and tested. In the latest management jargon, we need to move away from 'alternatives-focused thinking', in which GIS is seen as valuable simply because it offers something new and perhaps a little neater in terms of data management; and towards 'value-focused thinking', in which attention is centred on the benefits to the overall planning process that can result from new and imaginative applications to key problems.

CHAPTER 6

Health-care planning and analysis: the critical role of geographical modelling and information systems

6.1 INTRODUCTION

There has for a long time been a considerable interest in the geographical aspects of health-care. Much of the work undertaken by geographers has fallen under the banner of 'medical geography', with valuable contributions provided in four broad areas: spatial epidemiology, the spatial transmission of disease, the link between deprivation and disease, and the relationship between access and utilization. This literature is reviewed in Section 6.2. One feature that characterizes much of this tradition is that it is essentially descriptive, aiming to understand the relationships between observed patterns of, say, morbidity and social class. This is clearly an important and valuable task, and has added much to our knowledge of the structures and mechanisms that give rise to spatial patterns of inequality in health status, utilization, and resource provision. However, this chapter will argue and demonstrate that an additional dimension to the geographic analysis of health-care systems can be provided through the application and use of computer modelling methods, to add to the underlying theory and to generate a forecasting and prescriptive approach. The availability of such methods will be shown to be of crucial importance to health-service managers and planners, especially in the current climate of major changes in the management and delivery of health-care.

The chapter is structured as follows. First, it examines the variety of ways in which geography influences morbidity and health-care utilization, delivery, and finance. Second, it highlights the radical changes brought about through the splitting of the British National Health Service (NHS) into 'purchasers' and 'providers' of health-care, and the geographical concerns this raises. It proceeds to demonstrate how modelling methods can provide important insights in to health-care planning through the linkage of the geographical variations in health-care demand with the provision of health-care services in specific locations. The resultant patterns of utilization of health-care are important to understand, especially when changes to either the supply or demand side are being made. The chapter then reviews a number of different applications of geographical modelling methods to a variety of health-care systems. The final section summarizes the main points drawn out in the chapter and speculates on the future for model-based planning and analysis in the health service.

6.2 THE GEOGRAPHY OF HEALTH-CARE

6.2.1 Introduction

This section examines a variety of ways in which geography influences health-care – across the spectrum from health status and health-care utilization to resource allocation and management. Before that, however, it makes a couple of general observations concerning the NHS.

First, many commentators talk about the *creation* of the NHS in 1948. In fact the NHS was inherited from a disparate set of health-care systems ranging from private, employment-, and charity-based forms of provision (Leathard 1990). The hospitals and staff that made up these different health-care systems were transferred into the NHS in 1948. As there was no particular relationship between need and provision before 1948, it was only to be expected that post-1948 health-care resources would be unevenly spread across the country, with a notable concentration in London and the south east. In addition, resources were unevenly spread across cities and regions. This again is largely due to historical legacy. Since provision seems to influence patient referrals strongly (see Section 6.2.3), there will be inevitable geographical inequalities in access to health-care provision. The idea that access to resources is a function of where one lives has been labelled 'jurisdictional partitioning' (Pinch 1985). Attempts to rectify these in-built spatial inequalities have been ongoing over the last 40 years, and even now the process is far from complete. Formulae for redistributing resources between regions of the UK were introduced 20 years ago (the RAWP formula, discussed in Section 6.2.4). More recently, the focus has fallen on the distribution of facilities within metropolitan areas. The recent Tomlinson Report on the future of health-care provision in London is probably a precursor to many such studies around the country (HMSO 1992).

The second point relates to a fundamental misconception by the government when the NHS was established. It was thought that as health services would be free at the point of delivery and available to all irrespective of the ability to pay, more individuals, previously dissuaded from seeking care, would actually benefit from treatment and their health status would improve. A generally healthier population would then make fewer claims on the NHS and, after a period of relatively high funding, resources allocated could be cut back. As the Guillebaud Committee (HMSO 1956), set up to examine the costs of the NHS in the 1950s, demonstrated, the reverse was happening: there were increased demands for medical care. Behind this governmental misconception lies one of the main dilemmas of health-care provision: as more resources are made available, so the demand for service is greater. This has caused headaches to health-care planners ever since 1948. As publicly provided health-care budgets are finite, there has always had to be some kind of rationing. Often this has been accepted as a consequence of excess demand on the system, and 'tolerated' through allowing consultants to prioritize treatment as best they can. Delays in treatment are manifest through waiting lists, which generate substantial media attention and public concern. As we shall see in Section 6.3, the idea of rationing has come explicitly to the fore, with radical changes set out in the 'Working for Patients' White Paper (Department of Health 1989) and now implemented in England and Wales.

With these two general points in mind, the spatial dimension of health status and care can now be examined from a number of different points of view.

6.2.2 The spatial variation in health status: morbidity and mortality

The fact that health status varies across space is widely known and documented (see Eyles 1987, Haynes 1987, for general reviews). These variations apply at all spatial scales across the urban and regional hierarchy. The easiest to measure, and hence most commonly used, proxy indicator of ill health is mortality. Standard mortality ratios (SMRs) are calculated by comparing the actual number of mortalities in an area with the national average, taking into account age and sex variations in the area concerned. If an area was generating deaths at the national average, its SMR value would be exactly 100. Variations in standard mortality rates across the UK are shown in Figure 6.1. The areas with the highest rates tend to be in the north and west of the country, especially in the large conurbations. If these figures are broken down by regions, one can see as much variation within a region as across regions as a whole. Figure 6.2 shows standard mortality rates for males in Liverpool health authority. Again, it is possible to show that the more prosperous (outer) areas of Liverpool 'enjoy' lower SMRs, and vice versa. It is generally accepted that it is not the locality itself that generates high SMRs but the socio-economic characteristics of the households and the environment in which they live and work.

Measuring illness or morbidity is much more difficult, because most measures of morbidity are based on either institutional statistics (such as hospital admissions), or questionnaires (such as the Census). Admissions

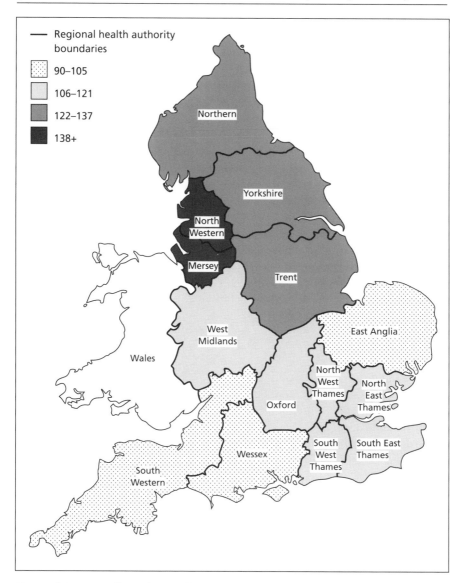

Figure 6.1 Age-adjusted mortality rates for all respiratory diseases 1992
(*Source*: adapted from data in *Regional trends* 1993)

statistics are clearly influenced by the levels of service provision, and questionnaires suffer from the reliability of the responses. There is also great difficulty in agreeing on and then measuring some objective definition of morbidity. For these reasons, mortality is often used as a surrogate for morbidity, given that legislation insists on the cause of death being entered on the death certificate. There is, at one level, an obvious relationship between death and illness; however, the majority of ill people are not in any

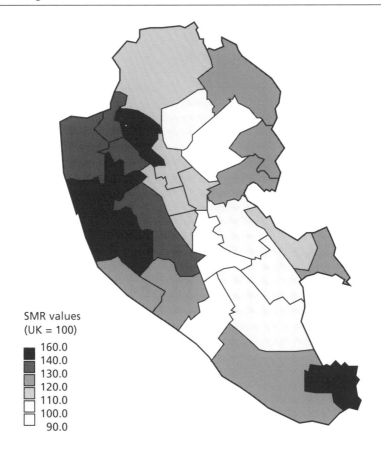

Figure 6.2 **Standard mortality rates for males across Liverpool**

immediate danger of dying, but suffer from conditions that prevent them from leading full and active lives.

A number of studies have attempted to examine illness through the approaches mentioned above. Eyles and Donovan (1990) looked at perceived illness in two contrasting boroughs of London: Tower Hamlets to the east characterized by high levels of deprivation, and Redbridge to the west, in a more leafy, green-belt suburb of the city. Table 6.1 shows the variations in response to questions concerning general morbidity. It is clear that residents in Tower Hamlets consistently scored worse than residents in Redbridge. There are always problems of reliability with 'perceived illness' in such questionnaire studies, but if people believe they are ill then that is an important indicator in its own right.

It is tempting to assign a rather simple causal model that explains this spatial variation: particular characteristics of 'place' are responsible for higher incidences of certain types of disease, which give rise to the high

Table 6.1 Self-reported morbidity in inner and outer London (*Source*: adapted from Eyles and Donovan 1990)

Item	Tower Hamlets	Redbridge
Tired all the time	18	12
Pain at night	11	6
Things getting me down	26	13
Feeling on edge	23	14
Trouble with stairs	16	10
Lose temper easily	23	16
Worry keeps me awake	12	4
I wake up depressed	12	5

observed mortality or morbidity ratios. Of course, the reality is much more complex than this, and a variety of explanations have been proposed, with deprivation and low socio-economic status figuring large as factors that contribute to greater ill health. These were the factors identified by the Black Report (HMSO 1980) as being significant variables in explaining variations in ill health, and form the bulk of the work of Townsend and colleagues in the north east of England (see Townsend and Davidson 1982, Townsend et al 1988). In addition, occupational structures are clearly important in explaining high death or illness rates in some towns or small communities. Table 6.2 shows standard mortality rates for a variety of different occupation types, and Table 6.3 for the standard social groups I–V. It is clear that manual workers (class IV and V) are much more susceptible to

Table 6.2 Standard mortality rates by occupation (*Source*: adapted from OPCS 1990)

Selected occupations	SMR
Armed forces	147
Labourers	141
Miners and quarrymen	144
Furnace and forge workers	122
Leather workers	114
Construction workers	111
Transport workers	111
Glass and ceramics	109
Warehousemen	108
Engineers	104
Clerical workers	99
Print workers	91
Salespersons	90
Professional workers	75
Admin. and managers	73
ALL	*100*

Table 6.3 SMRs by cause and social class (*Source*: OPCS 1990)

Causes	Class I	II	IV	V
Respiratory TB	32	35	128	279
Other TB	28	41	120	214
Neoplasms: stomach	50	67	127	158
Neoplasms: rectum	88	90	112	139
Chronic heart disease	66	71	122	154
Anaemias	55	61	117	134
Neoplasms: skin melanoma	133	126	89	82
Leukaemia	110	90	106	122
Neoplasms: brain	119	98	96	119

respiratory and heart diseases, whilst professional classes (class I and II) seem to suffer more from cancers (leukaemias), with particularly high rates of skin cancer. Some authors have gone on to argue that different socio-economic groups will also have very different lifestyles, leading to further variations in mortality and morbidity. Table 6.4 shows the variations across social groups in the propensity to take exercise and to smoke. If these variations exist nationally, there will be greater variations at the small-area level, where greater spatial concentrations of social groups exist. There are also references in the literature to variations in mortality rates by ethnicity (Marmot et al 1984), and investigations into the relationships between ill health and various environmental contaminations (Openshaw et al 1987, Lovett and Gatrell 1988).

Table 6.4 Smoking and exercise rates by social class (*Source*: *Social trends*, HMSO, London, 1984)
(a) Smoking

Class	Male (%)	Female (%)
I	17	15
II	29	29
III	40	37
IV	45	37
V	49	36

(b) Regular swimming sessions

Class	All (%)
I	18
II	15
III	14
IV	8
V	5

It might be suggested that people who fall into these ill-health categories tend to be spatially concentrated in 'inner-city' areas, and that this gives rise to the observed spatial patterns. There is the danger with this school of thought of falling into the ecological fallacy trap: that those people whose death results in high SMRs are deprived. Much current research is attempting to look at the deeper socio-cultural structures that generate attitudes towards and perceptions of health status and care, and how these might be reflected in different lifestyles.

The ability to analyse variations in mortality and morbidity (where such data exist) has been considerably aided by the arrival of widely available GIS packages. Many health authorities have been quick to realize the potential of GIS to store and retrieve spatial information at will. The mapping of incidences of illnesses, identified through utilization data, helps district health authorities (DHAs) to monitor trends in disease levels, and identify 'clusters' which may in turn help target campaigns of both preventive and treatment programmes. In 1986 CACI, a company specializing in the development of market analysis systems (see Section 4.3.5), was the first supplier to customize software towards the NHS, launching its PC-based Geographical Health Information System (GHIS). Since then there has been a great deal of collaboration between academia and health authorities, particularly through the RRL initiative: see the examples of the north-east RRL in searching for leukaemia clusters (Openshaw et al 1987) and in the mapping of various morbidities (Gatrell 1988), the north-west RRL for the mapping of spatial variations in health-care provision in Merseyside (Brown et al 1991), and the south-west RRL in mapping incidence of diseases in relation to population types (Wrigley et al 1988). Such GIS, based primarily on health-care data and geodemographics, also have an important role to play in marketing, especially for private sector hospitals keen to locate new developments in 'optimal' locations (Hale 1991). Once again, however, the use of GIS in this way is largely descriptive. The lack of suitable spatial analysis functionality means many of the strategic planning questions that are generated through descriptive work cannot be adequately addressed without looking to greater analytical power.

6.2.3 Spatial variation in health-care utilization

A basic tenet of the NHS Act of 1948 was that of equal access to health-care for all, irrespective of ability to pay, and this has remained the principle to this day. However, in practice there are substantial spatial variations in the rate at which health-care facilities are utilized. In one sense this is to be expected, as variations in morbidity are reflected in utilization of services. However, there is a consistency to the spatial variation in utilization rates that suggests that a more important cause than variations in morbidity is the availability and access to supply of health-care. In other words, everything

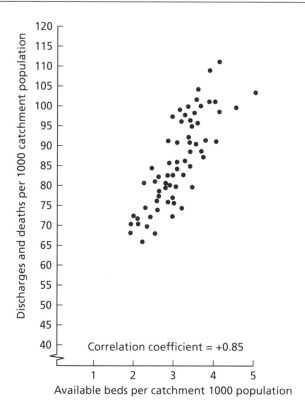

Figure 6.3 Number of deaths and discharges per head of catchment population
(*Source*: Mayhew 1982: 424)

else being equal, a person is more likely to become a hospital in-patient if he or she lives nearer to (or has better access to) a hospital (or a set of hospitals) than a person who lives further away. Figure 6.3 is taken from Mayhew (1982) and shows how, in London, the number of deaths and discharges per head of catchment population is strongly positively correlated with the number of beds per head of catchment population. It is, of course, possible to pitch the argument the other way: that health-care provision is located in the areas of greatest need. Whether this is the case or not is difficult to determine scientifically, but our feeling (and that of most observers) is that it is extremely unlikely that the *ad hoc* way in which hospitals have developed over the last century was somehow optimal in relation to need.

Work by the authors in North Humberside has shown that hospitalization rates by postal district, standardized for age and sex, for the main medical and surgical specialties are closely related to access to general hospitals, and that the range of variation from lowest to the highest is threefold (MVA/ULIS 1987). This is further confirmed by Figures 6.4 (Plate 6.1) and 6.5. The first shows standardized hospitalization rates for all acute specialties in west Yorkshire by postal sector. The location of the major acute hospitals has been overlaid. The visual correlation between high hospitalization rates and the location of the hospitals is very striking. Figure 6.5 repeats the exercise for the four Thames regional health authority (RHA) locations.

It is again tempting to propose that something intrinsically related to the presence of hospitals generates morbidity, which is then reflected in higher utilization rates. The reality is complex, and no satisfactory set of explanations has emerged to suggest why these spatial patterns exist. One immediate confounding factor is that many large general hospitals are located in central city locations, where the incidence of disease is higher and higher levels of deprivation exist.

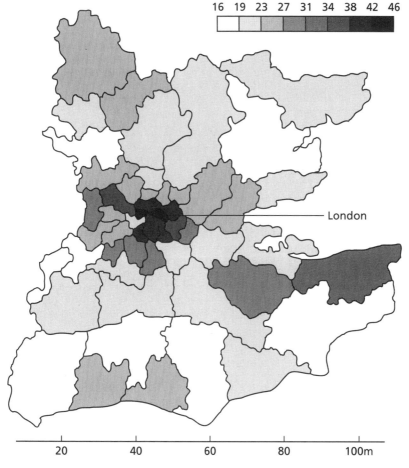

Figure 6.5 Hospitalization rates in the Thames RHA locations
(*Source*: GMAP 1994)

Since hospitalization rates are clearly also influenced by the actions of general practitioners (GPs), it is important to analyse GPs' referral patterns. A number of factors have been put forward to explain variations in such referral rates in different locations, such as practice size and characteristics of patients (Armstrong et al 1988, Wilkin and Smith 1987). Although there is not yet a firm explanation of GP referral behaviour, it is often suggested that if a GP's access to hospital facilities is poor then he or she will attempt to treat the medical episode within the community rather than refer. Conversely, a GP with good access to hospital care may more readily pass on the responsibility to a consultant. Inner-city GPs may also have longer list sizes and poorer diagnostic facilities, and perhaps be less aware of new forms of treatment (see Berkhout et al 1985 for a detailed analysis of referral patterns from GPs in Dewsbury). Taken together, these suggested causes may go some way towards explaining part of the spatial variation in utilization rates, but they hardly seem convincing as a total explanation. This area is therefore one which could benefit further from the attention of medical geographers.

6.2.4 Spatial variations in resource allocation

We have already noted that the health-care system that the NHS inherited in 1948 suffered from the legacy of a variety of *ad hoc* decisions by different types of organization on the location of hospitals and associated medical facilities. The original system of NHS funding, through the Regional Hospital Boards (RHBs), tended to reinforce these spatial inequalities. From 1948 to 1975 resources were allocated on a historical basis: RHBs received their last year's allocation plus whatever the general increase in funding for the NHS was that year. Not until the mid-1970s were steps taken to change this, with the establishment of the Resource Allocation Working Party (RAWP) and its recommendations for a new formula to allocate resources more equitably. The RAWP formula was based on a region's demographic profile, adjusted for local variations from average or national conditions. It was made up of the population of a region broken down by age and sex and further adjusted according to the standard mortality ratio of the region, the number of one-parent families (to try and capture 'deprivation'), and the amount of cross-boundary flow (there were extra resources for those regions with specialist facilities which might draw patients from outside their geographical boundaries). The RAWP (1976) report demonstrated the extent to which different regions of the country were being under- or over-funded according to this set of criteria. Using 100 as the expected level of funding, the range of actual funding varied from 85 (North West Region) to 114 (North East Thames Region). As a consequence, the Department of Health and Social Security (DHSS) accepted the RAWP recommendations and set

about introducing targets (the expected level of funding given the population structure and SMRs) that regions would move towards. Although there
has been a substantial volume of criticism directed towards the RAWP formula (see Bevan and Spencer 1984 for a review), it has achieved one major
objective: to bring about a regional redistribution of resources within the
NHS in England and Wales. By 1989 most regions were very close to their
RAWP target levels. This has not been without its pains, particularly in
London, where significant cuts in funding have been experienced.

At the more local, district level, many regions operated an allocation
system similar to RAWP although the problem was more complex because
of cross-boundary flows, through which some residents of one DHA might
receive treatment in another. This was a particular concern in London,
where many of the specialist teaching hospitals received patients on a
regional or even national basis. More recently, GIS has been used to help
target resources at the district levels. For example, the South East Thames
RHA have used CACI's ACORN geodemographic system to derive a set of
social deprivation scores, which could be used as a weighting factor for
allocating funds at the district level.

What has been the subject of little analysis is how health-care resources
are delivered to the different geographical areas of a district. Clearly, hospitals are located in a small number of discrete locations, whereas the demand
for health-care is distributed across a health authority. How well does the
delivery system meet those levels of need? For example, suppose each
Census ward in a city was a separate health authority. The RAWP formula
could be used to determine its resource allocation entitlement. Using health
authority data, it would be possible to compare the resource entitlement
with the actual consumption of health-care resources. Tong (1989) undertook this exercise for the two former health districts in Leeds, which
together constituted the Leeds Metropolitan District.

First, she took the 1981 Census data on population by age and sex for each
Census ward and combined this with the SMRs for each Census ward, using
the RAWP formula. This gives standardized populations, which can be
expressed as a ratio of the crude population. The next step was to obtain
data on the number of in-patients by specialty, generated by each Census
ward, irrespective of where they received treatment. For each in-patient the
national average specialty cost for treatment was used, and the total resource
consumption for each Census ward was calculated. The final step was to
compare the actual resource utilization with the RAWP expected utilization.
The results of this exercise are presented in Figure 6.6. The range of variation
is very dramatic – from less than 60 per cent of expected to over 170 per cent.
The interpretation of these results is difficult, but one interesting point
stands out. In the Otley Census ward, to the north west of the city, there is a
small district general hospital, which has been under the threat of closure for

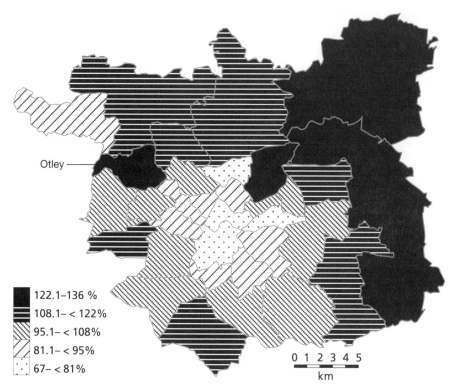

Otley

122.1–136 %
108.1– < 122%
95.1– < 108%
81.1– < 95%
67– < 81%

0 1 2 3 4 5
km

Figure 6.6   RAWP revenue: percentage to crude population (Leeds Census wards)
(*Source*: Tong 1989)

some time. The residents of Otley, a relatively prosperous market town and commuter suburb, have been vociferous in their objection to the hospital's closure. The fact that they are consuming substantially higher levels of health-care resources than RAWP would suggest they are entitled to may not be entirely unrelated to this. But it does provide a striking example, because it is not an inner-city area and probably no explanation could be offered purely in terms of morbidity.

In addition to RAWP, there has also been a great deal of concern and subsequent research in changing funding for GPs to account more adequately for higher incidences of visits (and hence treatment) in more deprived areas. Jarman (1983) first recommended deprivation payments, although it was not until the 1989 reforms that these became operational. The history and evaluation of such payments has been recently reviewed by Senior (1991).

The implementation of the 1989 UK government 'Working for Patients' White Paper has substantially altered the debate about funding in the NHS. However, the RAWP formula is still used for allocations at the regional level, although the formula at the heart of the allocation process is being reviewed once more with the objective of accounting for the effects of supply on utilization. The results of this review should be implemented in the 1994 allocation exercise.

6.2.5 Conclusions and implications

These various dimensions of spatial variation in health-care status, utilization, resource allocation, and organization have profound implications for the planning and management of health-care services. It is essential that methods of planning and analysis used to determine the form, level, and location of service and resource provision reflect the important geographical components underpinning the health-care system. In other words, the planning process should have an explicit geographical focus.

The next section discusses recent changes in the structure of the NHS, and Section 6.4 examines how the use of a set of key concepts based on spatial modelling can provide the enabling framework for this to be achieved.

6.3 RADICAL REFORMS IN THE NHS

The relationship between demand and supply in the NHS was fundamentally reformed by the 1989 'Working for Patients' Act. For the first time in its history, the NHS was split into 'purchasers' and 'providers' of health-care. Purchasers are made up of DHAs and 'fund-holding GP practices' (that is, those who opt to control their own budgets and satisfy government criteria on eligibility: they must have a minimum list size of 8000 patients). They purchase health-care on behalf of their resident population from providers, either hospital trusts that have elected to achieve self-governing status or directly managed units (DMUs) – hospitals that have not elected, or achieved, trust status and are still under direct DHA management. After the April 1994/95 round of applications there were 399 hospital trusts operating in England.

The aims of the reforms have ostensibly been to give patients more freedom of choice as to where they are treated, and to allow moneys to flow directly with patients as they receive care. Thus this directly removes the problem of resourcing cross-boundary flows, seen as a particularly difficult concept to measure through the RAWP formula. The internal market system works by purchasers and providers negotiating contracts to treat a set number of patient episodes for an agreed price. It is argued that the introduction of such a pricing mechanism will increase efficiency and effectiveness, as providers must ensure prices are competitive (thus perhaps ensuring savings are made when possible) and that quality of care remains high in order to win new contracts in the future. Hence hospitals have entered a new marketing environment, where they must shop around for contracts and keep abreast of the services offered by 'competing' hospitals.

Such reforms are becoming widespread in many countries with similar welfare systems to that of the UK, and it is not difficult to understand the analogies now being made between health-care provision and retailing in the private sector (Kearns and Barnett 1992). Similarly, there are many commentaries on the rhetoric of the reforms and the reality of what might happen (Spurgeon 1990, Mullen 1990, Harrison et al 1989, 1990, Harrison and Wistow 1991, Mohan 1995). The reforms certainly raise some fundamental questions relating to spatial equity.

The first and most direct consequence of the reforms is the changes in resourcing. With a simplified resource allocation procedure, it is clear that certain areas will see increases in their budgets at the expense of other regions. Mohan (1990) investigates these changes in some detail, concluding that rural Britain in general and the 'shire counties' of the south east in particular will make substantial gains. Other aspects of the reforms will appear as more subtle spatial issues. First it is argued that patients really have very little choice in the system, since contracts (and hence referrals) are made by the purchasing authorities and patients may have to follow money, rather than the other way round. Although an individual may be able to lobby the health authority for changes, patients are tied into contracts once they are made. If all the contracts made are local, problems of access are reduced. If, however, a contract is made with a provider some distance from the purchasing district (or GP), patients (and relatives and friends) may be tied into long journeys to receive care or make visits. This may also come about if a local hospital is 'full' (that is, it has made sufficient contracts to fill its beds) or its prices become too expensive for a particular purchaser: note the outcry when Alder Hey children's hospital in Liverpool put on an additional price tag of £7500 for those who wished to get faster treatment, thus forcing some Liverpool purchasers to look elsewhere. These accessibility issues are compounded if patients are fortunate enough to attend a GP surgery which has its own budget. It is argued that, through their GP's budget, they may be able to get preferential treatment or shorter waiting lists than those patients tied into longer-term contracts through the DHA. Again, there has been much evidence of this two-tier system in operation during the early years of the reforms, with one of the best-reported cases in Essex and Hertfordshire, where consultants were told to 'fast track' patients of budget-holding GPs in order to secure extra finances. Clearly those patients who live close to GP fund-holders (who tend to be away from inner-city areas) might benefit from simply being in the right place at the right time.

The reforms also lead to other worries which may have profound geographical consequences. First, since purchasers have to pay explicitly for each patient treated, there are concerns that the most expensive patients will be unwelcome. Since these will tend to be the elderly and disabled, this will be a problem wherever such groups are in greater concentrations. Second, what happens to providers that cannot compete in this new system? The rhetoric of the reforms is that they will have to find ways to reduce costs and hence be able to compete on price. The problems facing London hospitals, which have very high running costs partly as a result of their role as teaching institutions, have received much recent publicity. If they are forced to put prices up in order to claw back the revenues required to run such hospitals, they may not be able to secure contracts for more routine operations. This has already caused great concern and has been the subject of a major enquiry into the future of London's hospitals (HMSO 1992). It is clear from this report that major changes are under way. It recognized the problem of the withdrawal of patient flows from outside London as purchasers secured cheaper services more locally. It concluded that closures and mergers were necessary, suggesting eight out of the nine medical schools in London should merge into four, with substantial removal of beds in certain locations. The potential difficulties associated with implementing the Tomlinson Report continue at the time of writing.

A third concern is the promotion of private medicine in these reforms and in a number of tax perks made throughout the 1980s (see Johnson 1990). Again, the rhetoric is based on greater patient freedom to secure the level of care they can afford. This is increasingly tempting to many people as the private sector is promoted as having the best equipment, very few waiting lists and facilities which may border on high-class hotel standards. Yet access to private facilities is dependent on income and general good health. Insurance companies will not deal with the aged and infirm (unless they have sufficient wealth), and additional insurance premiums are beyond the means of many households. Couple this with the fact that the location of private hospitals is not uniformly spread across the UK, and it is clear to see the spatial implications of the privileged access to private medicine.

In the light of all these concerns, there is a great need to monitor the impacts of the purchaser–provider split and to offer support for more equitable planning. For purchasers, the shift from the management of health facilities to the determination and satisfaction of the health-care needs of their resident populations requires a better understanding of the geographical dimension of these needs and of health-care utilization and outcomes. In particular there is a need to address the following questions:

1 Where do our residents currently receive care?
2 How does the performance of current providers compare with planned objectives and at what costs?
3 How will the pattern of referrals have to change if proposed contracts are to be implemented?
4 How does the actual utilization of health-care services relate to the expected utilization by small geographical area?

5 How do health status variables relate to socio-economic and other variables by small area?

For providers of health-care, the central issue must be the better understanding of where their and competitor business is being generated, in the light of such geographical variations in demand. Again, a checklist could be drawn up as follows:

1 Where do our existing patients come from (by age, sex, specialty)?
2 What is the potential for increasing the workload by attracting patients currently referred elsewhere?
3 What are the implications of proposed contracts on the future viability of units and specialities within the organization?
4 Who are our main competitors?

The central argument of the remaining sections of this chapter is the need for a SDSS or IGIS to provide a framework for examining these and additional 'what if' questions in the context of the reformed NHS described above.

6.4 KEY CONCEPTS FOR HEALTH-CARE PLANNING

6.4.1 Introduction

A formal requirement for health authorities to prepare and publish strategic and operational plans has only been in place since the re-organization of the NHS in 1974. As a consequence, there has been a premium on staff with skills in the application of methodologies that would assist the planning process. This is even more the case with respect to the geographical aspects of planning and resource allocation. This section describes a number of key concepts relating to the health-care system that, taken together, begin to provide a framework for planning at the regional and district level.

The SDSS or IGIS we describe in this section can be considered as having four main elements:

1 an information system;
2 a method for estimating missing data such as morbidity and speciality costs;
3 a modelling system for producing 'what-if' forecasts;
4 a performance indicators package.

These elements have to be integrated together in an appropriate framework which allows, for example, the adjustment of databases and the creation of scenarios so that the 'what if' tests can be accomplished. This means that the information system has to be more than a data system: it has to contain model outputs and performance indicators as well as raw data, for example.

This kind of system has now be developed, and we shall give examples based on real applications. A version known as HIPS (Health Information and Planning System) has been tested in a number of contexts, such as East Yorkshire health authority (MVA/ULIS 1987), Dewsbury health authority, (Clarke and Spowage 1984), the Piemonte health authority in Italy (Clarke and Wilson 1984), and Bradford and Airedale health authorities. We shall describe the aims and outputs of the latest version of HIPS, now known as HIPPS (Health Information for Purchaser Planning System) below.

6.4.2 HIPPS Information System

6.4.2.1 Introduction

HIPPS was developed through collaboration between Bradford health authority, Bradford Family Health Service Authority (FHSA), a university project team, and GMAP, whose skills were employed to carry out the more technical software development driving the user interface. In addition, a version of HIPPS was customized for the entire west Yorkshire region on behalf of Yorkshire RHA.

As described earlier in this chapter, GIS have been generating a great deal of attention and interest in the health service in recent years. However, there is concern within the NHS that technology is driving the application, rather than the reverse. The NHS is certainly no stranger to accusations of this nature, given the difficulties experienced with the many expensive IT projects such as DISS, DISP, and other systems.

The HIPPS project followed the strategy for IGIS set out earlier in this book, and, through liaison with the representatives from Bradford health authority, a detailed specification document was produced for the IGIS. This was then passed to Bradford for comment. In this way, a continual dialogue was maintained and the product was shaped by the end-user as much as the provider.

6.4.2.2 Objectives

HIPPS was specifically designed to meet the needs of a purchasing authority. The type of information held in the system was guided by the type of analysis in which a purchaser might be interested. The main role of the DHAs as purchasers of care is to ensure that their residents are provided with appropriate levels of health-care, given available resources. As a consequence, DHAs will need to know in some detail the current pattern of health-care utilization by their residents, at a suitable disaggregation of site of treatment, speciality, etc. From this information we derived residence-based performance indicators such as hospitalization rates, notional bed provision, and notional expenditure per head. The system makes these available at the sub-district level (by postal sector or ED), since earlier research suggested that there is generally more variation within a DHA than between DHAs.

Purchasers might also be concerned with facility-based (provider) performance indicators, such as mean length of stay, turnover interval, waiting lists, cost per case, etc. These act as a guide when planning and negotiating contracts for care. Indeed, at the conclusion of the project, the Secretary of State for Health announced that such indicators of NHS Trust performance would be published (*Times*, 24 Feb 1993: 1–2). Measures of outcome are also necessary if purchasers are to attempt to tackle the problem of monitoring quality of care.

In the post-'Working for Patients' NHS, the role of the GP is likely to grow in significance. As the internal market evolves, it is probable that purchasing authorities will wish to monitor the activity of GPs, as the gatekeepers to health-care, through GP-based performance indicators such as referral rates and national budgets. There would also be a need to link this back to resident population with GP catchments and descriptions of practice lists.

Finally, purchasers might wish to monitor flows of in-patients, particularly to analyse extra-contractual referrals, and during the development of contracts.

6.4.2.3 Data

Little out-patient data were available at the time of the project (this is likely to change in the future), so that the system focused on in-patient episodes only. The main source of data for HIPPS was the former Korner Minimum data set on in-patient episodes. It contained a wealth of information on each in-patient episode. The fields we eventually made use of were:

- age/date of birth;
- unit post code;
- patient classification;
- patient category;
- referring GP code;
- date of decision to admit;
- date of admission;
- admission method;
- site of treatment;
- specialty/consultant code;
- destination on discharge.

Other data sources which were used included:

- FR11 – used to generate a 'hotel cost' for each occupied bed day at each site of treatment;
- FR12 – used to generate total expenditure by specialty for each site of treatment;
- SH3 – used to generate available bed days, occupied bed days, and total episodes by specialty for each site of treatment.

The FHSA provided details of each resident registered with a GP in each of the relevant FHSA areas. From this we were able to generate profiles of each GP's list and also, from the provided unit post code, GP catchments at ED level. Census data were incorporated at the ED, Census ward, and postal sector levels. These data provide important denominations for calculating performance indicators such as hospitalization rates.

Geographical boundary data came from three sources. First, Census ward boundaries were accessed from the digital boundary data database, DBD81, at the University of Manchester regional computing centre. Digitized post sectors were provided from GMAP's own sources. Finally, ED boundaries were generated around the ED centroids – from the Small Area Statistics Package (SASPAC) – using Thiessen polygons, constrained by the digitized Census and boundaries. We produced the polygons with a Thiessen polygon routine in ARC/INFO, and there was considerable further processing of the exported 'arcs'. Thiessen or proximal polygons are defined by drawing a boundary around a set of centroids such that the boundaries enclose all points closest to that centroid (Brassel and Reif 1979, Mark 1987, McCullagh and Ross 1980, Rhynsburger 1973). Other sources of data which were provided by the authorities included lists and addresses of hospitals and general practices, along with corresponding geocodes.

An important step in the development of the IGIS was to build up the database for all geographical resolutions. The crucial field of information is the geographical reference (the unit post code), which provides the link between the data sources. We assigned Ordnance Survey (OS) co-ordinates to unit post codes with the postal address file (PAF). We used these co-ordinates to map point locations, such as practices and hospitals, which have a unit post code. We defined GP catchments using the unit post code in the FHSA register, assigning the co-ordinates to EDs with a point-in-polygon routine. Effectively, as our ED polygons are Thiessen polygons, this assigned the co-ordinate to the nearest ED centroid. Similarly, we assigned each Korner episode to an ED with the unit post code and aggregated the EDs up to generate the ward data. Finally, we generated post-sector Census population data by assigning ED centroids to post sectors with another point-in-polygon routine.

6.4.2.4 Data quality issues

In general, the quality of the data was very good, though some comments should be made. For non-health data we experienced two problems with the ED centroids used to generate the Thiessen polygons. The first occurred in a couple of cases where the OS co-ordinate, given as the centroid point, was clearly inaccurate. This was identified, because the centroid point lay well outside the Census ward of which that ED was a member. We went back to the official ED boundary maps to identify those EDs and assign an estimated centroid as an OS co-ordinate. The second problem occurred where EDs shared the same centroid co-ordinate, usually due to the 100-metre resolution of the co-ordinates. In the majority of cases we were able to identify these EDs on the base maps, and offset the centroids to create polygons best describing the boundaries.

A rare problem occurred where the ED boundary could not be identified on the base maps; it turned out that these were 'hidden' EDs, and too small to be plotted on a map. In effect these EDs form islands within another ED. We simplified this problem and offset the centroids again, to generate two adjacent polygons.

Health-related data were perhaps surprisingly good. The paper data (FR11, FR12, and SH3) were complete except for the smallest sites of treatment. We received Korner data direct from the district's patient administration systems. These contained a variety of errors in the fields supplied. Problems with such data are well known; data may not be recorded on admission of the patient, there are commonly problems with post codes, or errors can creep in during the entry of the data into the systems. Another difficulty with Korner data is the non-uniformity of the speciality coding. Specialties are derived from the consultant managing the in-patient, and indeed for the Airedale data we received only the consultant code in the episode data, and a table linking consultants to specialty. The problem arises where large hospitals employ consultants in very narrow specialties, such as urology or endocrinology, which in smaller hospitals would be treated in blanket general surgical or general medical specialities.

A final comment regarding the Korner data concerns its content. We became aware that it would be useful to monitor and track tertiary referrals (that is, referrals from one hospital to another), and in principle the HIPPS system allows this, since episodes could be identified as originating from a secondary provider, or have as their outcome a tertiary referral. Unfortunately, the collaborating authorities had not yet developed a methodology to link and track these referrals.

In the case of the 461 314-record FHSA register, only 5619 records were discarded because the GP codes could not be identified, and 5 per cent could not be included in the catchment data where the post codes were not identified. There are, however, acknowledged problems with the FHSA register, and populations derived from it do not match the Census. This is mainly due to 'ghostings', when people do not remove their name from a register when they move area and there is a lag before re-registration. There is also a problem typically associated with minority ethnic groups. If a person's English is poor, it is sometimes not possible to identify his or her re-registration, as the spelling of his or her name may be incorrectly recorded. This type of problem might well occur more often in Bradford, with its high minority ethnic population.

In conclusion, the quality of our data was very good, and few Korner records were excluded as being entirely without value. We understand that more recent data has placed a high priority upon the accurate coding of post codes, which is of great importance to the GIS, and we assume data quality will improve even further in the future.

6.4.2.5 Use of the information system

There are a number of different organizations and functions within the NHS where HIPPS could assist. These include the following.

(i) The Family Health Service Authority (FHSA)

The FHSA have responsibility for the funding of GPs, monitoring their performance, and identifying issues of concern in the realm of primary health-care provision. As described earlier, they collate the FHSA register, which identifies all individuals registered with a given GP practice. It is thus possible to examine GP catchment areas – those small areas from which GPs generate their patients. This is of a particular importance in some areas, as GPs are entitled to additional payments if their practice is located in a 'deprived' Census ward. Deprivation is measured according to the Jarman Index (Jarman 1983), which is a composite measure accounting for a range of Census-based indicators. The crucial assumption is that the characteristics of the individuals on the register will be the same as those of the Census ward the practice is located in. With HIPPS it is possible to map out each practice's catchment area in terms of where individuals who are registered with that practice actually reside. The results of this exercise proved quite interesting. Figure 6.7 (Plate 6.2) shows the location of individuals registered with the Shipley Health Centre. As can be seen, the practice has a catchment area that is quite extensive, covering several Census wards, which means that patients are likely to come from a diverse range of socio-economic backgrounds. It would thus seem unfair to give additional payments to GPs whose practice just happened to be in a 'deprived' neighbourhood (the phenomenon of geographically dispersed catchment areas was common across Bradford GPs, irrespective of the type of location – city-centre, suburban, rural), and this calls into question the use of the existing method of providing additional funds to GP practices with under-privileged patients. At the present time, FHSAs are being combined with DHAs as joint purchasing authorities, but the essence of the argument still applies.

(ii) Public health medicine

The main responsibility of the public health division of a purchaser is to identify the health needs of the resident population by small area and to ascertain to what extent these needs are being met. In addition, health promotion and health gain have become in-patient functions since the publication of the 'Health of the Nation' White Paper (Department of Health 1992). HIPPS can assist the public health department in a number of ways. First, measures of morbidity can be generated to examine both the overall level of health-care demand in Bradford and how this varies geographically. Second, utilization of health-care services can also be examined. (Refer back to Figure 6.4 (Plate 6.1), which shows how hospitalization rates for general medicine vary by postal area in west Yorkshire.) It is interesting to observe the high correlation between hospitalization rates and hospital location. The key question is whether the areas of low hospitalization rates are more affluent, healthier places to live, or whether the presence of hospitals actually generates higher GP referral rates. The degree of variation in these rates should also be of concern to public health.

(iii) Planning in a DHA

The planning function within the DHA is concerned with developing arrangements for the provision of health-care services to Bradford residents in the future. This involves negotiating with health-care providers to determine agreed contracts for the provision of defined categories of treatment and care. A prime requirement in this respect is to understand existing relationships between small areas and providers: where do our residents currently receive care? HIPPS can provide this information in both tabular and mapped form for all specialisms and all providers by small area. Figure 6.8 (Plate 6.3) shows the hospitalization rates across west Yorkshire for cardiothoracic surgery, while Figures 6.9–6.11 (Plates 6.4–6.6) plot the origins of patients for this treatment at three different major surgical hospitals. The catchment areas are clearly demarcated and there is very little overlap. These flow patterns are interesting to study in their own right. However, they are particularly important to understand in the context of future planning, and the next step is to assess the impact of any proposed changes to contractual arrangements. This can be undertaken through the use of the modelling system within HIPPS that is described in the next section.

6.4.3 Development of the HIPPS modelling framework

6.4.3.1 Introduction

We feel that some form of predictive modelling capability is essential if purchasers are to come to terms successfully with the concept of 'managing the market'. It is not the purpose of this section to provide a detailed review of the history of modelling within the health service; for the background to modelling see Clarke (1984) and Mayhew (1986).

Many of the models developed during the 1980s used an approach known as 'microsimulation' (Clarke and Spowage 1984, Clarke and Wilson 1983, 1985b) within a comprehensive modelling framework. HIPPS adopts a similar comprehensive framework to that of the earlier applications, but uses an alternative technique, that of spatial interaction modelling. Such patient-flow modelling began with work in the Operations Research Service of the DHSS in the late 1970s and early 1980s. The success of the development and application of these methods led to some RHAs applying flow models to strategic planning in the mid-1980s. The limitations of these approaches meant they were not well received and in many cases flow modelling came to a halt. Further attempts were made by the West Midlands RHA as part of a strategic planning exercise in 1988, with more encouraging results, and a simple gravity model was used during the RAWP review process in 1987/88. An earlier Trent project, and a recent GMAP project to model patient flows in the four Thames regions, support the application of spatial interaction models in the health service.

We believe that it will considerably enhance the accuracy and utility of such models to include the intermediate stage in the flow of an in-patient to hospital, namely the GP referral. The understanding of the referral process is an important step in producing accurate predictive models of patient flows (Berkhout et al 1985). As we have already mentioned, the role of GPs has changed in the post- 'Working for Patients' NHS; they are likely to become more involved either in contractual issues as fund-holders, or in influencing the content of DHA contracts. If purchasers are to 'manage the market' they must be able to predict the influence of the GP (King and Newton 1990). There has been much empirical research into GP referral patterns (Morrell et al 1971, Crombie and Fleming 1988, Day and Klein 1991), and many attempts to explain the variation in GP referral rates (Armstrong et al 1988, Berkeley 1976, Wijkel 1986, Wilkin and Smith 1987). Many factors have been suggested which might explain referral rates: practice size, accessibility of facilities, characteristics of patients and GPs, and so on. Despite the many studies there is not yet a firm explanation of GP referral behaviour. Some studies either failed to produce conclusive evidence as to which factors are associated with variations in referral behaviour, or warn of drawing firm conclusions (Marinker et al 1988, Moore and Roland 1989, Roland 1988, Roland et al 1990, Wilkin et al 1989). With no consensus of opinion from these behavioural approaches, our mathematical modelling of the GP referral provides an important alternative for investigating the relationships. If conclusive results do come from behavioural studies, such as a doctor's utility index (Healey and Ryan 1992), then these results may be incorporated into our model at a later date.

Essentially there is a series of stages in the model process, described here in turn. The first is to model the registrations of patients within EDs to practices and to simulate the practices' catchment population. Most research concerning general practice concentrates on referral rates, though studies have been made regarding spatial patterns of registration to practices and the effects of geography upon the utilization of primary health facilities (Hays et al 1990, Knox 1978, 1979, Phillips 1979). Also, a recent study by Martin and Williams (1992) used a spatial interaction model to study the general practice registration of residents in the Bristol area.

We have used the same type of spatial interaction as in the study by Martin and Williams, a production-constrained spatial interaction model as described in Chapter 3. However, whereas that study used point origin locations to generate market areas for GPs, ours modelled the flows of residents of EDs in the study areas to practices. The ED populations are the smallest spatial resolution of population which might then be projected forwards to generate new model scenarios.

The model itself may be written as:

$$S_{ij} = A_i\, O_i\, W_j\, e^{-\beta_j c_{ij}} \tag{6.1}$$

where:

i is the origin ED;
j is the destination practice;
S_{ij} is the flow of residents registering at practice j from i;
O_i is the demand (the population of the ED i);
W_j is the attractiveness of practice j (list size);
c_{ij} is a travel-cost variable (straight-line distance from the ED centroid to the practice);
β_j is a destination-specific distance-decay parameter;

A_i is a balancing factor defined as:

$$A_i = 1/\sum_j W_j\, e^{-\beta_j c_{ij}} \tag{6.2}$$

to ensure that patient flows sum to the known patient totals at the origin end.

The model produces a matrix of predictions of flows between the EDs and practices. In the application, 908 EDs and 98 practices were modelled.

An observed flow matrix of 908 by 98 was extracted from the FHSA register. Predicted flows are generated by altering the input values to the variables. If the model is run using the observed values of these variables, it follows that the predicted flows should match the observed flows. In reality this is not the case, and the process of calibrating models must be introduced to look at the similarity between observed and predicted flows. Calibration will be considered later.

Running this first model gives two useful products. First, total predicted practice list size can be generated.

$$S_j = \sum_i S_{ij} \tag{6.3}$$

This may be used in the generation of performance indicators.

Secondly, we have the matrix of flows from ED to practice $\{S_{ij}\}$. This is interesting to study in its own right; for instance, what happens if a GP leaves a particular practice? We have modelled the GP attractiveness as a simple function of the number of GPs and mean list size for each practice. By altering the number of GPs in practices, new scenarios may be generated to look at the impacts of these changes.

The flow of patients from ED to practice is also used as the input to the next stage of the model. This uses a number of probabilities, generated from the observed data, to predict the number of in-patients by specialty arising in each practice (that is, it predicts secondary care morbidity by specialty). Throughout the remainder of this section, this part of the model will be referred to) as the *morbidity model*. This can be written as:

$$S_{ij}^{as} = S_{ij} A_j^{as} \tag{6.4}$$

$$T_{ij}^{las} = S_{ij}^{as} M_j^{as} B_l^{as} \tag{6.5}$$

where:

a is age split;
s is sex split;
l is specialty;

A_j^{as} is the probability of a patient in practice *j* being aged *a* and sex *s*;

S_{ij}^{as} is the breakdown by age and sex of the practice list S_{ij};

M_j^{as} is the probability of a patient registered to *j* of age *a*, sex *s*, requiring an in-patient episode;

B_l^{as} is the probability that an in-patient referral by age *a* and and sex *s* is to specialty *l*;

T_{ij}^{las} is the referred patients from *i* through *j* to specialty *l* by age *a* and sex *s*.

This can be aggregated to generate:

$$T_{ij}^{l} = \sum_a \sum_s T_{ij}^{las} \tag{6.6}$$

the flow of patients from ED to practice who are then referred on for treatment in specialty l. Also:

$$T_i^{l} = \sum_j \sum_a \sum_s T_{ij}^{las} \tag{6.7}$$

and

$$T_j^{l} = \sum_i \sum_a \sum_s T_{ij}^{las} \tag{6.8}$$

are used to generate residential and practice performance indicators respectively. This concludes the morbidity model and the first half of the overall model, the primary care model. At this point the action of the general practitioner has been simulated and demand for health-care generated.

The second half of the model is a group of production-constrained models to determine which hospitals each practice refers the in-patients to for treatment in a given speciality. For this secondary care model, 19 production-constrained spatial interaction models have been calibrated. For each speciality l = 1–19. The models are written as:

$$T_{jk}^{l} = A_j^{l} T_j^{l} W_k^{l} e^{-\beta_k^{l} c_{jk}} \tag{6.9}$$

where:

j is the practice;

k is the provider unit (hospital);

l is the specialty of treatment;

T_{jk}^{l} is the flow of in-patients from practice j to hospital k, calibrated for specialty l;

T_j^{l} is the demand (in-patients from practice j for specialty l), generated above;

W_k^{l} is the attractiveness of hospital k for specialty l (observed episodes);

c_{jk} is the travel-cost variable (straight-line distance);

β_k^{l} is the destination-specific distance-decay parameter for hospital k, calibrated for specialty l;

A_j^{l} is the balancing factor to ensure that the flow sum matches known totals, defined as:

$$A_j^{l} = 1 / \sum_k W_k^{l} e^{-\beta_k^{l} c_{jk}} \tag{6.10}$$

T_j^{kl} represents the flow of referrals by practice j to hospital k for treatment in specialty l.

The referral patterns can then be displayed. We modelled the attractiveness of the hospitals simply as a function of the observed episodes treated, multiplied by a factor initially set at unity (which can be altered to create a scenario where hospital case load is increased or decreased by a percentage factor). The impacts of changes in the provision of hospital services may be analysed. T_j^{kl} may also be aggregated:

$$T_k^l = \sum_j T_{jk}^1 \tag{6.11}$$

which may be used to generate facility performance indicators.

6.4.3.2 Model calibration

From the above description it should be clear that there were in total 19 production-constrained spatial interaction models to be calibrated. The only parameter in the models to be calibrated are the destination-specific distance-decay parameters, β_j and β_k^l. The values are assigned to these parameters by calibrating the models to observed data, using maximum likelihood (Batty and Mackie 1972, Knudsen and Fotheringham 1986). We assigned an initial estimated value to beta in the models and produced a program to run the model iteratively, generating new values for beta using the maximum likelihood equation until the observed mean travel distance equalled the predicted mean travel distance. After each iteration, goodness-of-fit statistics were generated for the flows, such as the sum of squared deviations, R-squared, and standardized root mean square error (srmse).

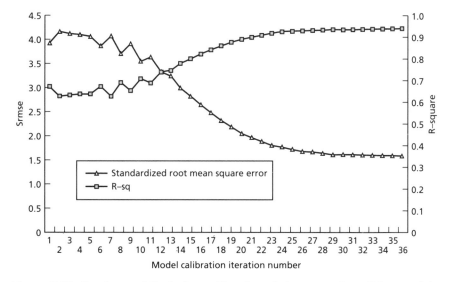

Figure 6.12 Goodness of fit during calibration of the general medicine model
(*Source*: GMAP 1994)

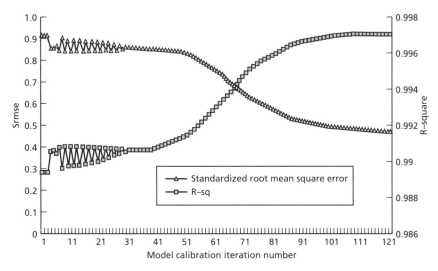

Figure 6.13 Goodness of fit during calibration of the medical oncology model
(*Source*: GMAP 1994)

Having reached a satisfactory convergence between observed and predicted mean distance travelled, the beta values were then saved for use as input in later baseline or scenario models.

Typically, convergence occurred within 40 iterations (see Figure 6.12 for a plot of goodness-of-fit statistics for general medicine), and this can be compared with the plot of statistics for the medical oncology model (see Figure 6.13), which incurred many more iterations.

6.4.3.3 Results of the model

The result of the model calibration process still does not produce a perfect match between observed and predicted flows. The overall R-square is 0.809, which is a reasonable fit. However, this does mask a variation within the goodness of fit of flows to each destination (practice) of between 0.994 and 0.318. It is interesting to note that this reveals a problem, within this first model, of dual site practices: that is, practices whose GPs provide services from more than one surgery in different locations. To illustrate this point, Figure 6.14 shows the observed registrations of residents to practice B83637, at 2 Chatham Street, Bradford. This is the practice with the poorest fit and it is clear from the map that registrations are based upon two foci. To solve this problem with the data would improve the performance of the model.

Table 6.5 summarizes the overall fit of the models by specialty, sorted in descending order of R-square. It shows that much better fits were achieved at this level of resolution. This is partly due to the small number of destinations in several circumstances. For instance, perfect fits had to be achieved

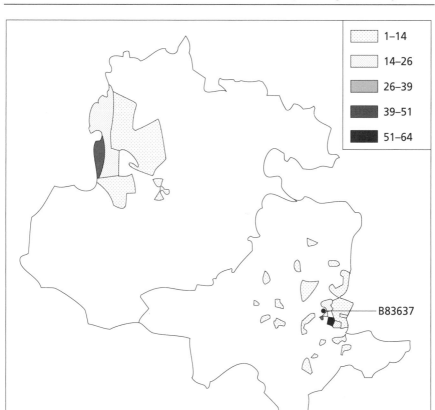

Figure 6.14 Patient registrations for practice B83637
(*Source*: GMAP 1994)

for the four specialties with only one provider. Excluding the single provider models, which were only calibrated in order for the impact of a new competing provider to be modelled, the fit remains very good with R-square of 0.925 or above, with the exception of the bucket speciality group 999, which produced the worst fit.

To summarize, the calibration procedure seems successfully to have produced models of a suitable goodness of fit, whose use in practice will be described in the next section.

6.4.3.4 An example of the use of the model

The first stage in use was to run the baseline model for comparative purposes. The model was then run for two simple scenarios, and produced output in the

Table 6.5 Calibrated practice to provider model results (*Source*: GMAP 1994)

Specialty	Competing providers	Observed mean distance	Predicted mean distance	Distance variance	Sum of squares	Standardized root mean square error	R-square
1	1	492	492	0.00	0.00	0.000	1.000
2	1	491	491	0.00	0.00	0.000	1.000
3	1	480	480	0.00	0.00	0.000	1.000
4	1	368	368	0.00	0.00	0.000	1.000
5	1	422	422	0.00	0.00	0.000	1.000
6	2	345	340	5.25	1 532.93	0.324	0.998
7	2	403	403	0.71	195.03	0.486	0.997
8	2	404	398	5.65	53.49	0.463	0.997
9	2	390	382	8.02	1 982.19	0.568	0.996
10	2	447	438	9.17	232.01	0.857	0.990
11	2	466	458	7.99	2 211.35	1.070	0.975
12	5	370	382	11.71	107 193.97	1.130	0.972
13	3	403	399	4.62	1 200.28	1.079	0.969
14	3	410	398	11.49	88 908.12	1.122	0.967
15	3	558	550	8.36	50 144.77	1.358	0.955
16	3	428	413	15.06	154 754.47	1.593	0.934
17	10	484	465	18.67	18 227.90	1.415	0.926
18	4	353	354	1.03	118 961.37	1.428	0.925
19	9	429	404	25.81	8 152.14	2.011	0.821
Maximum	10	558	550	25.81	154 754.47	2.011	1.000
Minimum	1	345	340	0.00	0.00	0.000	0.821
Mean	3	429	423	7.03	29 144.74	0.784	0.970
Median	2	422	404	5.65	1 532.93	0.857	0.990

form of tables and maps. The tables were generated in Microsoft Excel, while the maps were produced with GMAP mapping software.

The first model focused upon simple changes in GP practice size. The model scenario reduced the number of practitioners in B83021, Keighley Health Centre, from seven to three. Table 6.6 shows the observed list size of the practice, and other practices situated in the locality, which are affected by flow deflections. The baseline model appears to work well, though slightly over-estimating list size. Comparison of Figures 6.15 and 6.16 reveal that the baseline model has replicated the geographical patterns of registrations, though the baseline is slightly more centralized. The scenario model drastically reduces the list size of the practice with the local practices

Table 6.6 Deflection rates following reduction in practice size (*Source*: GMAP 1994)

Practice code	Observed list size	Baseline		Model 1		Difference between obs and pred
		List size	Practitioners	List size	Practitioners	
B83006	9 107	10 614	6	12 288	6	1 674
B83021	13 471	14 621	7	7 298	3	−7 323
B83023	7 084	8 716	4	10 137	4	1 421
B83033	12 169	12 693	6	14 821	6	2 128
B83047	2 478	3 182	1	3 684	1	502

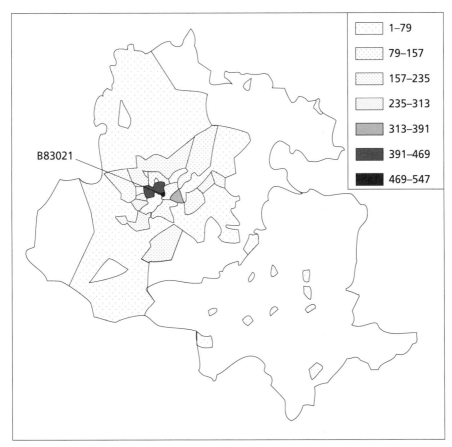

Figure 6.15 Observed set of practice registrations for centre B83021
(*Source*: GMAP 1994)

picking up the bulk of the deflected flows. Figure 6.17 illustrates the change in geographical patterns of registration, predicted by the model. With the exception of all flows being of a lower magnitude, the changes from such a small alteration in the scenario are too subtle really to stand out.

Table 6.7 Number of episodes: baseline and Model 1 (*Source*: GMAP 1994)

Practice code	Specialty code	Baseline referrals	Model 1 referrals	Difference between obs and pred
B83008	100	119	137	18
B83021	100	156	78	−78
B83023	100	101	118	17
B83033	100	127	148	21
B83047	100	42	49	7

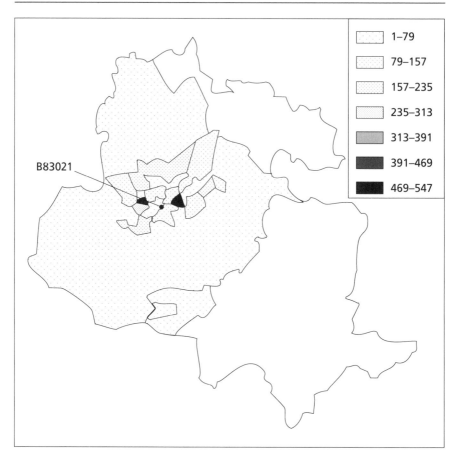

Figure 6.16 Modelled set of practice registrations for centre B83021
(*Source*: GMAP 1994)

Table 6.7 displays the observed, baseline, and predicted referrals by the practices. Referral rates are constant, as none of the parameters associated with the morbidity model in the scenario was altered. As one might expect, the change from baseline to predicted referrals replicates the pattern seen in the change in list size.

The scenario in Model 2 looks at another possible query. The attractiveness of Airedale General's surgical specialty is increased to reflect greater bed availability, while St Luke's and BRI, and other acute general surgical providers, remain constant. Increasing the attractiveness of Airedale General in Model 2 increases the episodes treated at Airedale at the expense of the other providers. These providers are exactly as one would expect given the spatial arrangement of patients and hospital locations.

Figure 6.18a shows the pattern of practices referring to Airedale General.

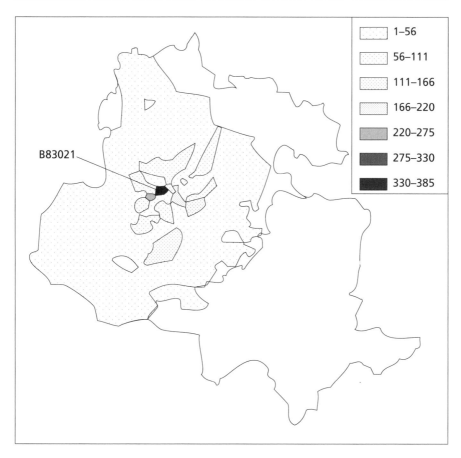

**Figure 6.17   New set of practice registrations for centre B83021 following
               reduction in size**
(*Source*: GMAP 1994)

The effect of Model 2, shown to the right of Figure 6.18b, is that Airedale now captures flows of patients previously going to the Bradford hospitals.

The third example that we present relates to an ever-more important problem faced by the NHS. There is an increasing tendency towards the development of large, specialized facilities in a smaller number of hospital locations. This typically requires a reduction in the number of existing units and the assessment of the impacts of this process of rationalization and concentration. On the one hand, concentration of facilities should lead to the development of 'centres of excellence' and increasingly more effective patient care. On the other hand, it reduces the overall level of access to these facilities and therefore potentially raises important issues of spatial equity. In Section 6.4.2.5 we examined the issue of the provision of cardiothoracic surgery in west Yorkshire. We noted that there were substantial

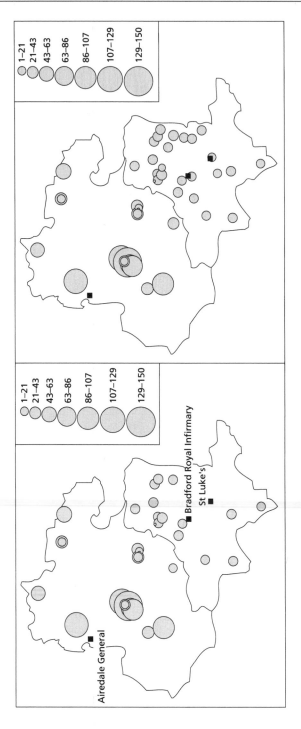

Figure 6.18 **(a)** GP practice referrals to Airedale General for general surgery
(b) GP practice referrals to Airedale General for general surgery
following increased attractiveness

(*Source:* GMAP 1994)

degrees of variation in hospitalization rates for this specialism across the Census wards of west Yorkshire and a very strong relationship between high hospitalization rates and facility location. At the time of writing there is a proposal to reduce the number of hospitals performing cardiothoracic surgery from three to two. This involves the closure of Killingbeck Hospital in east Leeds and its incorporation with the existing surgical unit at the Leeds General Infirmary in the centre of the city. The other remaining facility would be the Bradford Royal Infirmary.

We have used the modelling framework described above to attempt to measure the impact of this change and to assess the impacts of alternative proposals on hospitalization rates within west Yorkshire. Figure 6.19 (Plate 6.7) demonstrates that the model predicts that a move to the Leeds General Infirmary would result in a substantial reduction in hospitalization rates in the east of the city of Leeds coupled with an increase in the west of the city (the traditional catchment area of the Leeds General Infirmary). However, this would do little to redress the significant imbalances in hospitalization rates that exist in the south west of the region, particularly in the areas around Halifax and Huddersfield (see again Figure 6.8 (Plate 6.3)). We therefore examined the impacts of relocating the Killingbeck facility to the Huddersfield Royal Infirmary. Figure 6.20 (Plate 6.8) illustrates the impacts on hospitalization rates as predicted by the model. We see a significant reduction in hospitalization rates in the Leeds region (where they were already high), but a marked increase in the Huddersfield and Halifax areas, which currently have very low hospitalization rates. From an equity perspective this solution would probably be deemed preferable to the merger of Killingbeck and the Leeds General Infirmary. However, we suspect that arguments about efficiency are deemed more important than equity in the current environment of planning within the NHS. We would argue that through the use of this type of geographical analysis the impacts surrounding various options can be explored in some detail, and that this should lead to better-informed decisions being made in cases such as this.

These three simple examples of the model serve to illustrate how it might be used to examine the implications of changes to primary or secondary facility provision. Multiple changes can be included in each scenario, and entire facilities can be opened or closed. Other parameters, such as residential population and the morbidity variables, could be changed in the scenario to simulate changes in the structure of the population and referral behaviour. Bed availability and mean length of stay can be altered according to the changes in provision by the acute units. From all these changes, performance indicators are generated to enable sophisticated analysis of the health authorities' potential decisions, and alternatives may be assessed.

6.5 CONCLUDING COMMENTS

We hope that we have illustrated the crucial contribution of geographical analysis to health-care planning and management. While medical geography has a strong and respectable tradition, it has perhaps failed to contribute to the decision-making process in the NHS to the extent it should have. One obvious reason for this is the lack of recognition in the NHS of the role geography plays in a variety of different aspects of health-care. Stronger lobbying by geographers is needed here.

It is also important to set out a research agenda for applied medical geography. Clearly, this will be wide ranging, but in the context of the issues discussed above a number of areas warrant detailed attention. First, there is the current review of health-care organization that was mentioned briefly above. This gives hospitals the option of 'opting out' from health authority management and charging the health authority for services provided. GPs will also have the option of being allocated budgets in such a way as to purchase services for their patients from different hospitals, which will compete for patient referrals – the so-called internal markets. Although there has been a great deal of response and criticism from observers to these proposals, there has been little analysis of how they will affect the geography of health-care.

Secondly, there is the important issue of continuing the tradition in medical geography of spatial epidemiology (see the excellent review of Thomas 1992). In particular, there is an urgent need to monitor and forecast the spatial distribution of individuals who are HIV-positive or have full-blown AIDS. Each AIDS case is estimated to cost the NHS some £28 000, and at the local level the difference between 20 and 50 cases per year has serious resource implications. There have been few attempts to forecast these numbers at the regional or district level, and those that have been made tend to be based on trend extrapolation. For such an important issue there is a need to apply more sophisticated modelling technology. Clarke and Rees (1989) have proposed an approach based on microsimulation methodology that would explicitly focus on the process of viral transmission between different sexual orientation groups and drug abusers, with a high level of spatial representation (see also Smyth and Thomas 1994). Medical geographers argue forcefully that AIDS is an important subject for analysis, and there is therefore a need to demonstrate the contribution that can be made.

A third area relates to the promotion of GIS. There exists considerable scope for the development of GIS within health-care systems, and it is important to build on the progress reported in this chapter.

There is no shortage of interesting research issues in quantitative medical geography. What remains problematic is the setting of priorities and being seen both to contribute to important contemporary debate and to perform analysis that will inform and assist decision makers at a number of levels. The challenge exists, and time will tell whether it is responded to.

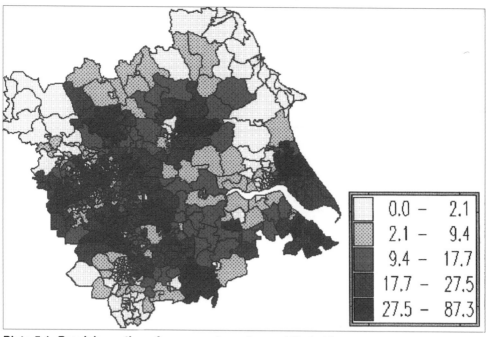

Plate 5.1 Provision ratios of grocery stores in west Yorkshire

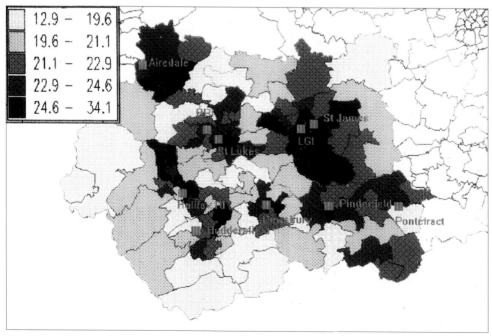

Plate 6.1 Hospitalization rates in west Yorkshire

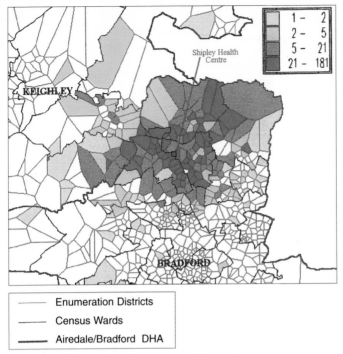

	1 – 2
	2 – 5
	5 – 21
	21 – 181

Shipley Health
Centre

KEIGHLEY

BRADFORD

——— Enumeration Districts

——— Census Wards

——— Airedale/Bradford DHA

Plate 6.2 Shipley Health Centre: GP catchment area

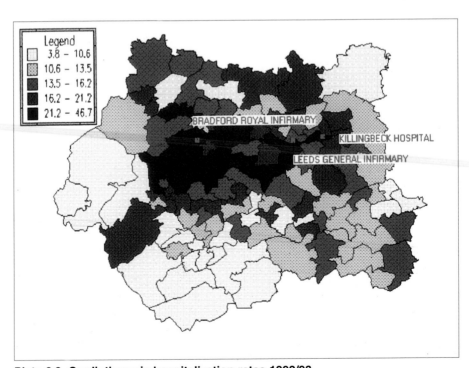

Legend
3.8 – 10.6
10.6 – 13.5
13.5 – 16.2
16.2 – 21.2
21.2 – 46.7

BRADFORD ROYAL INFIRMARY

KILLINGBECK HOSPITAL

LEEDS GENERAL INFIRMARY

Plate 6.3 Cardiothoracic hospitalization rates 1992/93

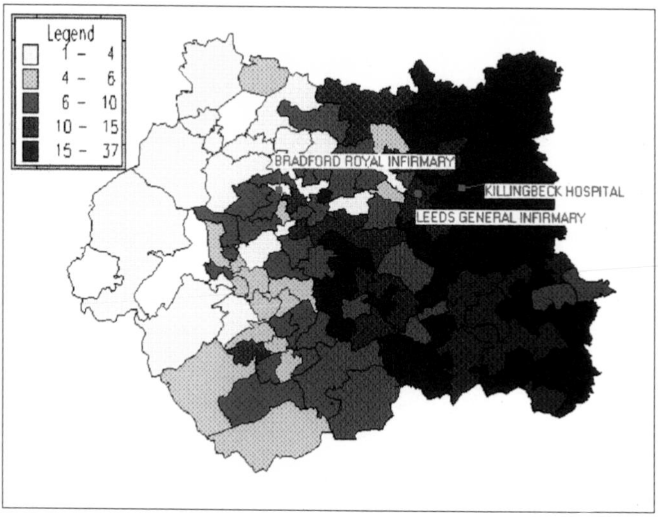

**Plate 6.4
Cardiothoracic patient
flows to Killingbeck
Hospital 1992/93**

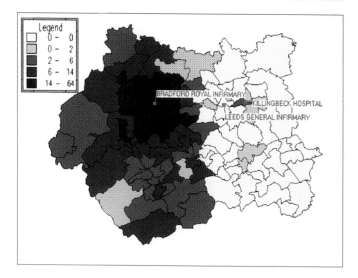

**Plate 6.5
Cardiothoracic patient
flows to Bradford Royal
Infirmary 1992/93**

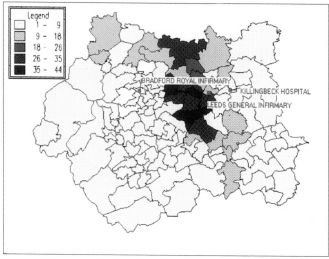

**Plate 6.6
Cardiothoracic patient
flows to Leeds General
Infirmary**

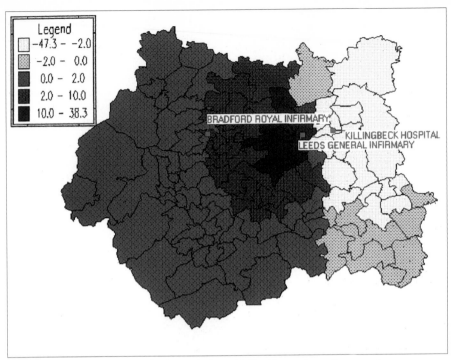

Plate 6.7 Changes in cardiothoracic hospitalization rates after moving Killingbeck facilities to Leeds General Infirmary

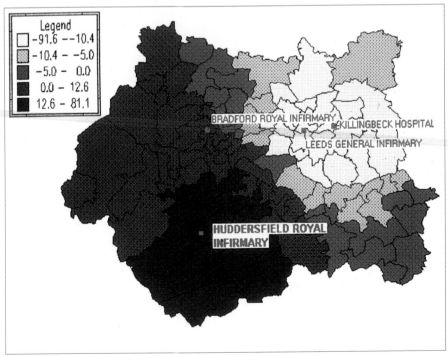

Plate 6.8 Changes in cardiothoracic hospitalization rates after moving Killingbeck facilities to Huddersfield Royal Infirmary

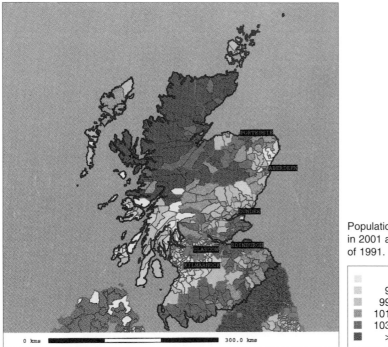

Population all age groups
in 2001 as a percentage
of 1991.

	< 98
	98-99
	99-101
	101-103
	103-105
	> 105

Plate 7.1 Population change by postal district 1991-2001: Scotland

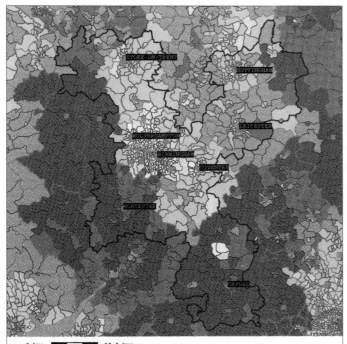

Population all age groups
in 2001 as a percentage
of 1991.

	< 99
	99-103
	103-105
	105-107
	107-109
	> 109

Plate 7.2 Population change by postal district 1991-2001: central England

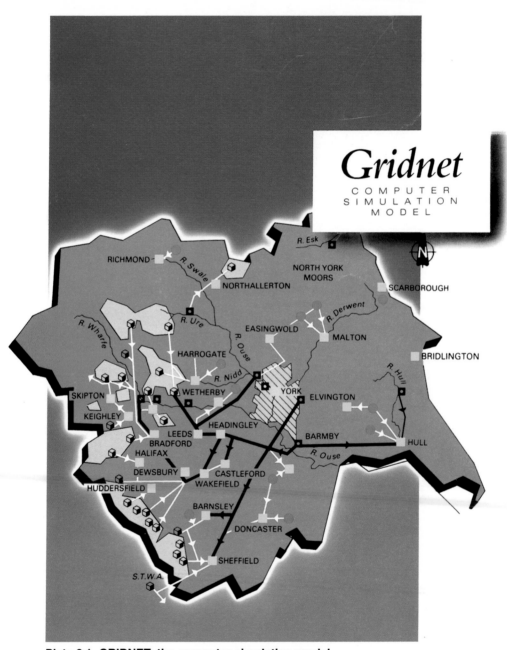

Plate 8.1 GRIDNET: the computer simulation model

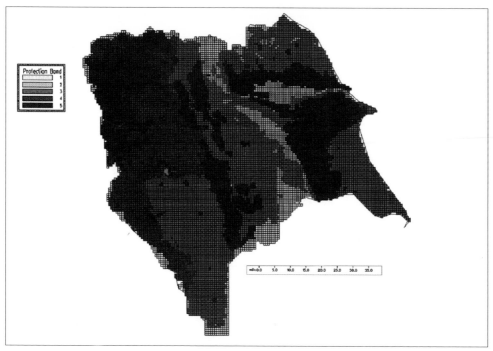

Plate 8.2 Borehole protection bandings

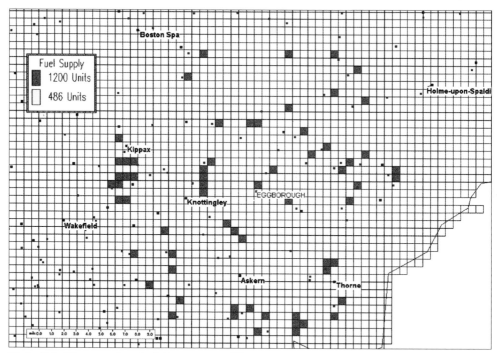

Plate 8.3 Biomass inputs into new gassifier

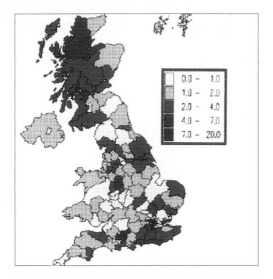

Plate 9.1a Market penetration for Fiat in the UK

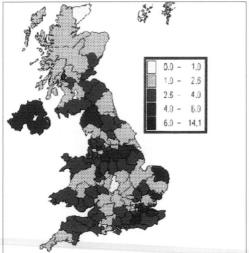

Plate 9.1b Market penetration for Toyota in the UK

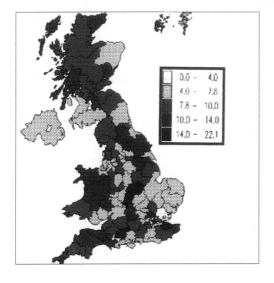

Plate 9.1c Market penetration for Peugeot in the UK

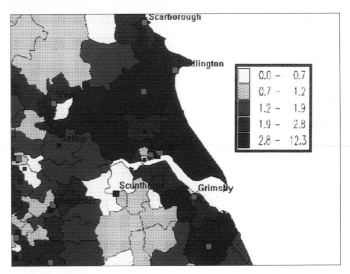

Plate 9.2a
Market penetration for
Fiat in Humberside

	0.0 – 0.7
	0.7 – 1.2
	1.2 – 1.9
	1.9 – 2.8
	2.8 – 12.3

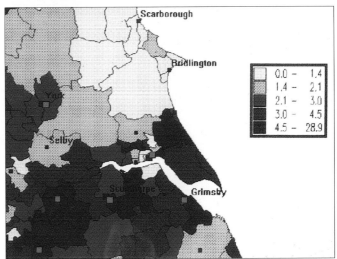

Plate 9.2b
Market penetration for
Toyota in Humberside

	0.0 – 1.4
	1.4 – 2.1
	2.1 – 3.0
	3.0 – 4.5
	4.5 – 28.9

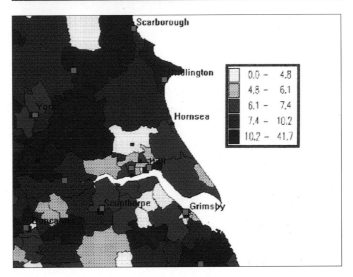

Plate 9.2c
Market penetration for
Peugeot in Humberside

	0.0 – 4.8
	4.8 – 6.1
	6.1 – 7.4
	7.4 – 10.2
	10.2 – 41.7

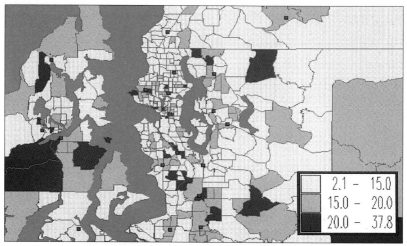

Plate 9.3a Ford market share in Seattle/Tacoma, USA

	2.1 – 15.0
	15.0 – 20.0
	20.0 – 37.8

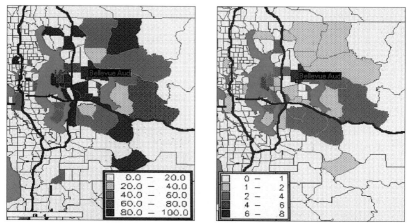

Plate 9.3b Sales at Bellevue Audi as a percentage of Audi's market

	0.0 – 20.0
	20.0 – 40.0
	40.0 – 60.0
	60.0 – 80.0
	80.0 – 100.0

	0 – 1
	1 – 2
	2 – 4
	4 – 6
	6 – 8

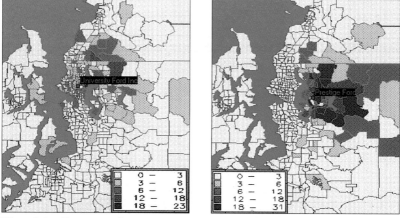

Plate 9.3c Patterns of sales in city-centre and suburban locations

	0 – 3
	3 – 6
	6 – 12
	12 – 18
	18 – 23

	0 – 3
	3 – 6
	6 – 12
	12 – 18
	18 – 31

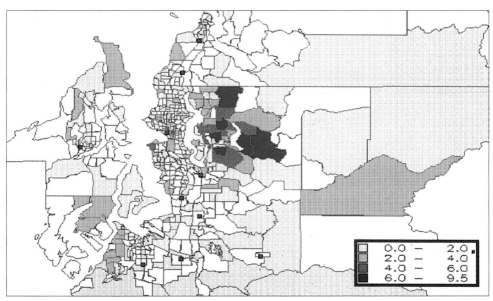

Plate 9.4a Opportunity for additional Pontiac sales

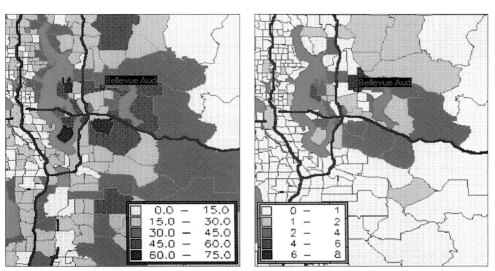

Plate 9.4b Number of households with income greater than $60 000 alongside Audi sales

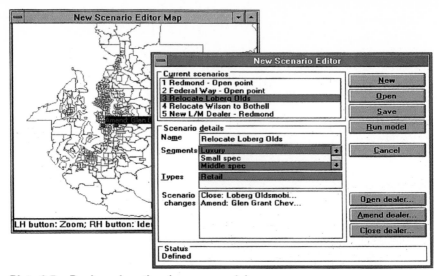

Plate 9.5a Dealer relocation (new scenario)

Options					Options			
Census Tract	Base CHE All All Reg	RlctL CHE All All Reg	Base OLD All All Reg	RlctL OLD All All Reg	Census Tract	Base GlnGr CHE All C Or	RlctL GlnGr CHE All C Or	RlctL GlnGr OLD All C Or
1.00	16.91	16.91	0.73	0.66	1.00	0.63	0.63	0.09
2.00	23.64	23.64	1.74	1.62	2.00	0.91	0.91	0.18
3.00	7.52	7.52	0.68	0.61	3.00	0.32	0.32	0.07
4.00	29.03	29.03	1.86	1.70	4.00	0.81	0.81	0.25
5.00	8.55	8.55	1.44	1.32	5.00	0.36	0.36	0.19
6.00	19.65	19.65	1.70	1.54	6.00	0.83	0.83	0.22
7.00	10.01	10.01	0.74	0.67	7.00	0.32	0.32	0.09
8.00	5.77	5.77	0.48	0.43	8.00	0.11	0.11	0.07
9.00	5.56	5.56	0.53	0.49	9.00	0.18	0.18	0.06
10.00	4.67	4.67	0.39	0.36	10.00	0.11	0.11	0.04
Total	7429.28	7429.28	1044.80	935.47	Total	433.19	431.47	118.21
Mean	15.35	15.35	2.16	1.93	Mean	0.90	0.89	0.24

Plate 9.5b Results screen showing loss of Oldsmobile sales

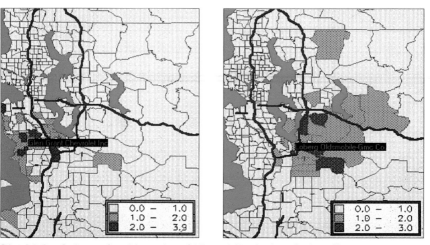

Plate 9.5c Sales gained by other Oldsmobile dealers in locality

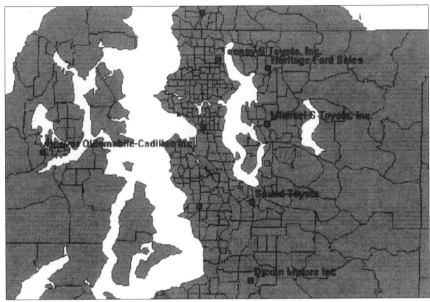

Plate 9.6a Toyota representation in Seattle and new optimal locations

Dealers	Sales
Bayside Toyota, Inc.	1048
Heritage Ford Sales	955
Michael-S Toyota, Inc.	829
Burien Toyota	655
Sound Toyota	655
247.00	576
291.00	492
Tenney-S Toyota, Inc.	441
Toyota of Puyallup Inc	432
Titus-Will Ford Sales Inc	411
Rodland Toyota, Inc.	365
Aurora Toyota	356
Doxon Motors Inc	316
Mc Carroll Toyota	143
Hoover Oldsmobile-Cadillac Inc	106
Total	**7780.00**
Mean	**518.67**

Plate 9.6b New sales forecasts for two new Toyota dealerships

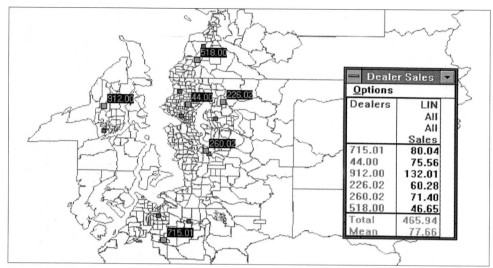

Plate 9.7a Optimal locations of new Lincoln dealerships

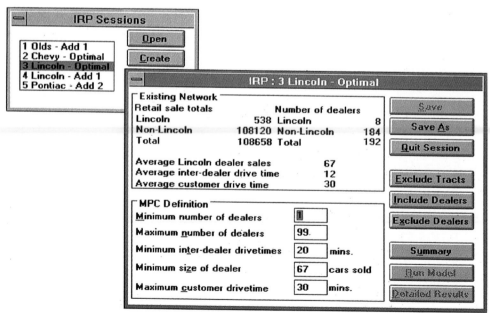

Plate 9.7b Optimization selections (new scenario)

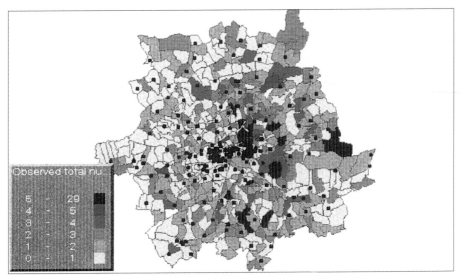

Plate 10.1 The Network effect for a major bank in London

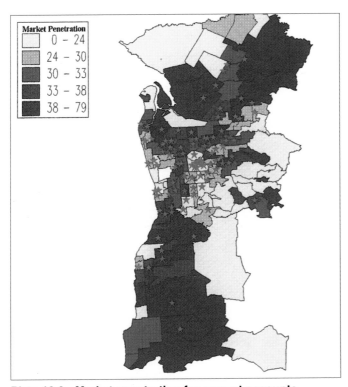

Plate 10.2 Market penetration for current accounts

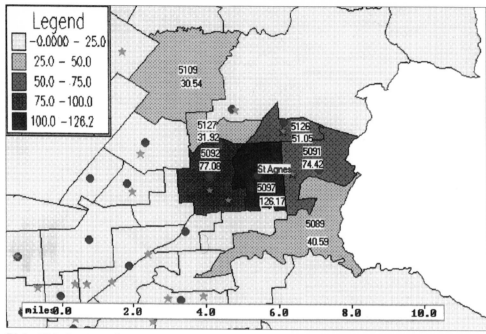

Plate 10.3 Increase in new transactions following new branch opening in St Agnes, Adelaide

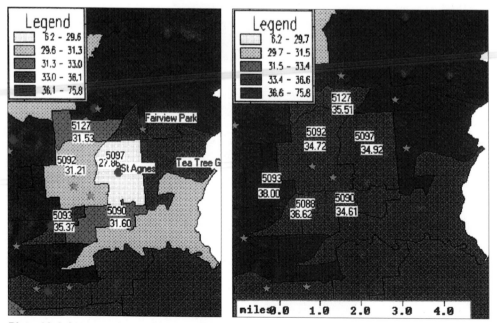

Plate 10.4 Increase in market penetrations following new branch opening in St Agnes, Adelaide

CHAPTER 7

Model-based GIS for urban planning

7.1 INTRODUCTION

Urban or town planning has a long history in the UK and other developed countries. Since the development of local government in the nineteenth century, the planning of local-based services has been the responsibility of elected local authorities. Urban planning today covers a wide range of activities, from development control through to enhancing local economic activity. We shall adopt a wide-ranging definition of urban planning that includes activities such as social services, housing, education, and labour-market planning as opposed to the rather narrow definition implied by the responsibilities of a local authority planning department. The range of responsibilities of a typical local authority is outlined in Figure 7.1.

In addition, local authorities have to interface with a wide variety of other agencies, such as central government departments (including the DTI), health authorities, local education authorities, and private sector organizations.

One of the major challenges in urban planning is balancing the spatial equity of facility or service provision with that of economic efficiency. Most residents would prefer high levels of access to schools, libraries, employment offices, etc., but it is usually not economic to provide such ubiquity of outlets. Hence, important decisions have to be made on what are the best locations for facilities given variations in local need or demand. Similarly, providing adequate transport infrastructure must be balanced against the environmental impacts of new roads. As we shall describe, it is to support such decision making that GIS can be of value in the urban planning field.

Each of the responsibilities shown in Figure 7.1 could be a major section of this book in its own right. Indeed, the literature on GIS and transport, for example, is itself now very comprehensive. The aim of this chapter is to highlight the power of GIS and modelling through a number of illustrative examples rather than to provide a comprehensive review of spatial analysis in all aspects of urban planning.

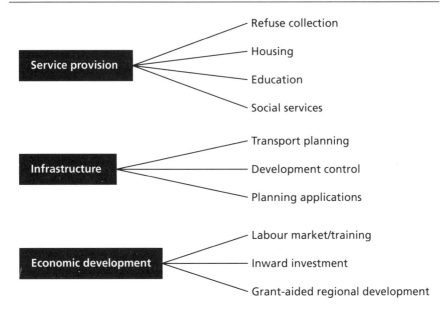

Service provision
- Refuse collection
- Housing
- Education
- Social services

Infrastructure
- Transport planning
- Development control
- Planning applications

Economic development
- Labour market/training
- Inward investment
- Grant-aided regional development

Figure 7.1 Local authority responsibilities

7.2 CHANGING INFORMATION REQUIREMENTS IN URBAN PLANNING

The task of strategic planning by local authorities has been neatly summarized by a number of authors as 'wicked' (Mason and Matriff 1981, Bruton and Nicholson 1987). The characteristics which make for wickedness are 'interconnectivity', 'complexity', 'uncertainty', 'ambiguity', 'conflict', and 'societal constraints':

> In particular, difficulties are raised by the inter-relationships between specific proposals for land use and development and wider social and economic issues, at both the same 'local' level and at higher more general levels in the hierarchy, and between specific local development and resource allocation and politics.
>
> (Bruton and Nicholson 1987: 56)

The complexity of many planning issues has been one of the primary reasons for the involvement of academics, particularly through the introduction of spatial modelling methods. This level of partnership and collaboration between academia and the planning profession has waxed and waned over the last 30 years or so. Faith was perhaps greatest in the 1950s and 1960s, when structure plan design was synonymous in many Western countries with the systems analysis route to planning (Chadwick 1971, McLoughlin 1973). The subsequent dissatisfaction with and decline of modelling methods is a story well told elsewhere (Batty 1989, Bertuglia et al

1994b). Batty (1989) concludes his review of models in urban planning by speculating that the future success of analytical methods may revolve around advances in computer graphics. Indeed, the arrival of GIS seems to have offered a potential new era for spatial analysis in local authorities and for collaboration with academia. The 'seduction' caused by computer graphics has surely played a large part in this turnaround. The impact of GIS in urban planning will be explored below.

Although technological advances in GIS have been important for putting spatial analysis back on the planning agenda, it is important to emphasize the new conditions of planning which have altered the traditional role of land-use planning in many Western countries. The nature of local government in the US and most Western European countries has undergone rapid change in the late 1980s and early 1990s. We have seen a general shift away from the direct provision of services towards 'enabling', so that the role of local government in the 1990s is perceived as being the provider of a strategic management framework for service provision. This may take place either privately, or through competition between the public and the private sector. Barnekov et al (1989) label this the 'pursuit of the private city' and describe the transatlantic transfer of the ideas of 'privatism', which originated in the US.

In the UK this new era of privatism is illustrated by far-reaching changes in the provision of social services (following the Griffiths Report (HMSO 1984)), moves towards local financial management of schools together with greater freedom of choice (the Education Reform Act of 1988), and fundamental – though not, it now appears, irreversible – change in the system of local government finance and taxation (the ill-fated community charge or 'poll tax'). Key urban development policies have also been contracted out to the private sector in the UK (Barnekov et al 1989, Robson 1988), with the development of 'urban development corporations'.

In addition, we have witnessed the increased enthusiasm of central government for regulating local government expenditure (for example, through charge capping). This implies that local authorities must be able to prove that they are using resources effectively. This in turn involves setting objectives and targets for services, monitoring of spending, and performance evaluation. This concern with value for money extends to specific council wards in the UK, as council tax payers and local councillors are increasingly concerned with what items of service are actually being utilized or 'consumed' by charge payers. Similarly, in the US and Canada, there are growing consumer pressures on local authorities to become more proactive in managing urban growth and the local services required to support it (Newkirk 1991).

Thus, the 1980s' trend towards growing consumerism in service markets has fed through to local authorities. Many are now developing marketing strategies to identify target groups for services, and advocating a greater role for research and development. The UK Local Government Management Board (LGMB 1990) identifies a number of broad areas where further research is most likely. These include:

1 research to identify and measure key environmental, social and economic problems;
2 research to identify 'priority' groups for the more effective targeting of resources;
3 the interviewing of local residents to obtain the perception of how well served residents believe themselves to be;
4 the identification of gaps in service provision and subsequent recommendations of new services;
5 research to monitor effectiveness and efficiency of service provision;
6 research to develop a framework for strategic management based on the mission statements of different local authorities.

In all these areas there are clear requirements for accurate and timely local information.

In addition to the demands for better information within local authorities there is also an increasing demand for research which may help local authorities win extra resources available from outside the region, often in direct competition with other regions. In a sense, cities or regions have always been in competition, especially for grants available for community development projects (Loney 1983) and urban development programmes (see Barnekov et al 1989, Robson 1988, for a fuller list). However, this battle for resources has never been greater than at the present time, with new EC grants on offer and additional opportunities available in the UK for inward investment as the government seeks to move administrative departments from London to the regions. We shall see examples of this type of research in later sections (see also Ashworth 1990 on the need to market cities more effectively for inward investment and tourism).

These statements of research priorities, coupled with the changes in the pattern of local government finance, the growth of accountability, and the increasing focus on monitoring and evaluation of services (rather than direct provision) all point to a need for better information to guide a more rational planning process. In order to perform these functions, local authorities will need to develop new information sources; make more co-ordinated use of information; and develop more effective ways of handling, analysing, and presenting information. The obvious way in which to achieve these objectives is through an appropriate information systems strategy (cf. Gault and Peutherer 1990).

7.3 GIS AND LOCAL AUTHORITIES: THE STORY SO FAR

The rise of GIS in local authorities has been very closely linked with the development of digital mapping and automated cartography. This is the conclusion of a large number of studies which have reviewed GIS usage in local government in different areas of the world (for instance, Bromley and Coulson 1989, Campbell 1994, Worrall 1989, 1990b, Couclelis 1991, French and Wiggins 1991, Kim 1990, Yeh 1990). These studies describe the principal use of GIS as supporting the production and maintenance of land-use maps (see Markham et al 1990). It is not surprising that local authorities have concentrated first on land-use maps given the demands for local land searches, which, if co-ordinated manually, produce a 'time consuming repetition of a laborious information co-ordinating process' (Aspinall and Boxer 1993). In addition, there are obvious financial gains from computerizing the production of basic maps. Buxton (1989), in an evaluation of GIS in Cardiff, estimates an annual saving of £165 000 through the introduction of GIS for automated map production.

In addition to land-use information systems there are a number of impressive socio-economic planning information systems constructed through the combination of public-domain and in-house spatial data sets. Mahoney (1989) estimates that 80 per cent of all data held by local authorities has a geographical dimension. In the UK this is well illustrated by Worrall's work in Telford (Worrall 1989, 1990b, Worrall and Rao 1991), where he has produced an impressive amalgamation of data to support information dissemination. Similarly, Gault and Peutherer (1990) outline the Strathclyde Regional Council small-area information system, which shows the degree to which data are available at a variety of different geographical scales.

It is evident that these 'descriptive' information systems dominate the literature on GIS in local authority use, and there is little widespread evidence of more analytical functionality for manipulating and linking these data sets. However, there are notable exceptions to this. Boalt and Bernow (1991) describe the STOCAB information system set up in Stockholm, which not only contains a vast array of different spatial data sets but has also been linked to statistical packages such as SAS, and transport-route allocation models. This enables planners to conduct factor analysis to measure and monitor social differentiation within the urban region. In Waterloo in Canada, Newkirk (1987, 1991) describes WATGUM, a modelling system coupled to GIS to allow planners to test and evaluate the consequences of urban simulation exercises on the basis of a set of decision rules which test for 'acceptable' outcomes. Other examples from around the world include Roy and Snickars (1993) in Perth and Helsinki, Garner (1990) in Australia, and Wood (1990) in Tacoma in the United States. Section 7.5 outlines the importance of linking spatial models to GIS.

7.4 EVALUATION OF CURRENT GIS IN URBAN PLANNING

In Section 7.2 we identified a need for urban information systems to assist in the planning process. The typical, and natural, reaction of local authorities has been to invest in an 'off-the-shelf' GIS solution which produces the type of information system described in Section 7.3. Thus, a specification of the needs of an authority is drawn up, and a range of GIS vendors is invited to tender for the contract to supply an appropriate system. The vendors are obliged to show how effectively their existing software product can be applied to fit the needs identified. The contract is then awarded on the basis of the price of the system and its functional suitability for the job in question.

There are three objections to this type of procedure. The first is that it can often lead to fragmentation within an organization. The nature of local authorities is such that it will often be an individual department, such as planning or highways and engineering, which presses hardest for the introduction of GIS (and which may also ultimately finance the system) simply because different responsibilities are associated with different forms of local government. As a number of authors point out (Mahoney 1989, Lopez and John 1993), it may be hard to maintain a balanced view of corporate goals in this situation. However, the facility to integrate existing data sources across an organization is one of the crucial arguments for having a GIS in the first place. There is evidence that as experience with GIS grows then so does the level of corporate interdependence. Swainston (1993) describes the centralized production of maps to serve Dudley's planning organization as a whole, whilst Aspinall and Boxer (1993) report the corporate use of GIS in Wakefield council for the investigation of land-use charges.

A second objection to this style of approach is that it is technology-led (see also Worrall 1989). Thus, typically, a GIS will be concerned to fit the series of problems defined by the planning agency into the box provided by an existing system. An even worse situation can arise, perhaps after a GIS has been purchased by the organization, when the availability of technology actually starts to influence perceived planning or policy goals. The alternative here is to build customized solutions tailored to specific problems. This might involve the development of a specifically local-authority-orientated GIS package; or better still, a customized approach for individual authorities using fundamental GIS building blocks, such as a database management system combined with a mapping or graphics capability. The latter approach is demonstrated in the next section.

The third objection is that existing GIS provide data systems rather than information for policy making. Typically, existing systems can synthesize and integrate spatial data effectively, but do not have the capacity to forecast or plan ahead, or provide the basis for impact ('what if?') analysis. This sort of analysis seems to have slipped from favour in recent times, perhaps because, as Breheny (1987) suggests, local authority planning has never tended to be very proactive, more often relying on marginal changes which do not move far from the status quo. Coupled with the difficulty of obtaining corporate co-operation on GIS development as a whole, this helps to explain why GIS has been associated with low-order planning tasks, not high-order strategic planning activities (see also Worrall 1989, Worrall and Rao 1991, Couclelis 1991). Yet, as Harris (1991a: 5) emphasizes: 'If we lack the ability to choose between alternative plans by comparing their outcomes, then planning as the premeditation of action has a very weak basis indeed.' Harris (1991b: 4) continues by stressing the importance of suitable methods for 'planning support', methods which must include modelling approaches to 'support the planner by making part of his [*sic*] task automatic through the use of models to flesh out a plan and follow through its consequences'.

It would not be entirely fair to proceed to the next section without considering the fact that the types of modelling approach which we are now advocating have themselves been subjected to many criticisms in the past, as alluded to in Chapter 1. Ottens (1990) identifies five restraining features in the application of model-based methods in planning:

1 reluctance to adopt an objective analytical approach in planning;
2 data-hungry models;
3 computational inefficiency (especially graphics);
4 lack of training of planners in automated information processing;
5 large investments required for hardware and software.

To these, Breheny (1987) adds the indictment that modellers themselves have, in the past, been excessively technically orientated, and concerned more with the methods themselves than with their application. This is perhaps especially the case with many of the models produced within the 'regional science school', with their all too frequently large and complex array of variables and assumptions (Bolton 1985, Newkirk 1991, Openshaw 1986b). Finally, one could also add that customized solutions to information systems design can be financially expensive, even when compared to the considerable sums required for conventional GIS development.

These kinds of reason can be used to explain the relatively low uptake of modelling methods into planning in the 1970s and early 1980s (cf. Batty 1989). However, there is evidence that many of these conditions are now breaking down. The use of computers is now widespread and no longer the domain of the technically minded. At the same time, the power of computation (in terms of storage, processing capability, and graphical capabilities) has continued to develop at an astonishing rate. As far as 'data hungriness' is concerned, we are seeing increased data capture; for example, through the Census, through government departments and within local government itself. Lastly, the kinds of institutional development outlined above are, perhaps, beginning to imply that the time for urban modelling has finally arrived.

The next section reviews more fully the role of IGIS in a number of areas of local planning. It argues that traditional (proprietary) GIS *can* produce useful information and analysis to aid policy formulation. However, it also argues that a second tier or level of analysis could be incorporated by coupling spatial modelling techniques with GIS. This message is now being echoed elsewhere in the literature on urban planning. Scholten and Padding (1991) label this the 'missing link' and advocate a 'conceptual model' of an ideal GIS which incorporates statistical methods as well as forecasting models. Similar arguments are made by Harris (1989), Couclelis (1991), and Brail (1990).

7.5 APPLICATIONS OF GIS AND SPATIAL MODELLING

The last section argued that the application of 'off-the-shelf' GIS packages to urban planning has been highly imperfect. We believe there is much to be gained through the incorporation of more explicit, analytical (specifically, model-based) techniques. This section illustrates applications of this strategy in four policy contexts: deprivation studies, retail site analysis, labour-market analysis, and population forecasting. Section 7.6 says something about comprehensive approaches.

7.5.1 Deprivation studies

The renewed interest in local authority accountability has placed issues of resource allocation firmly back on the political agenda. Local councillors are keen to make sure their locality receives the resources its population base requires, whilst urban or rural planners battle to quantify the levels of deprivation within their boundaries in order to try and win government or EC resources in the form of various deprivation relief grants.

GIS is increasingly seen as a tool which can aid the process of measuring and monitoring deprivation, and a number of studies now appear in the literature. There is a renewed interest in defining and measuring indicators for improving the targeting of resources on particular client groups within the community (see Clarke and Wilson 1994a for a history of such indicator studies). Worrall and Rao (1991) have exploited the overlay functions in GIS to produce a local need index based on the linkage between a number of Census variables (Figure 7.2). Similarly, Hirschfield et al (1991) have produced a useful set of simple indicators to evaluate the allocation of resources from the Urban Programme Grant in St Helens, Merseyside, UK. They overlay expenditure from the grant (by Census ward) with long-term

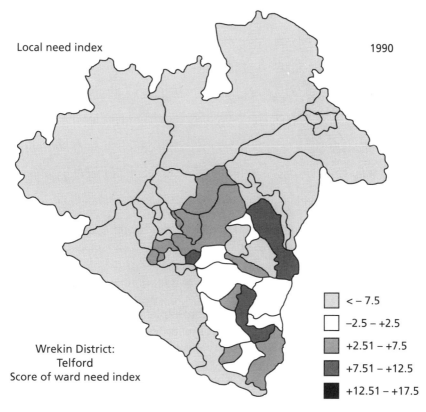

Figure 7.2 Local need index
(*Source*: Worral and Rao 1991: 147)

unemployment to produce a composite deprivation index. Boalt and Bernow (1991) use factor analysis to define a set of indicators to isolate and monitor change in socially disadvantaged areas (cf. Giggs 1970). Such indicator studies are generally presented through traditional choropleth maps. However, a number of studies have begun to experiment with redefinitions of polygon boundaries according to the actual distribution of households within those polygons (Higgs and Wright 1991), or by using 'population surfaces' based on the actual population distributions rather than whole ED polygons, many of which may contain large areas of zero population (Bracken 1991, Martin 1991, 1995).

These indicator studies provide a useful starting point in terms of deprivation measurement, but suffer from the traditional problem with such indicators: they are based on single zones and fail to take account of zonal interdependencies. Clarke and Wilson (1994b) articulate a new set of model-based performance indicators based on the principles of spatial interaction, and a number of these are measured for Leeds in Birkin et al

Table 7.1 Sets of performance indicators

Residence-based:
  Household incomes
  Quality of housing and residential environment
  Work opportunities
  Take-up of marketed goods and services
  Take-up of public goods and services
  Transport (generalized costs)

Organization-based:
  Efficiency of production
  Role in the pattern of provision
  Role in the labour market

(1994). The arguments are based on defining residence-based indicators which relate to individuals and households, and facility-based indicators which relate to the efficiency and effectiveness of organizations which serve residences. The indicators are broken down into sets, as shown in Table 7.1. What is clearly required is the incorporation of such indicators within an IGIS.

There have been other attempts to look at deprivation issues using GIS technology. Martin et al (1992) produced a novel study of gainers and losers in the recent changes to local taxation in the UK. For a small area of inner Cardiff, they looked at the number of households that were forced to pay more tax when the system changed from a property-based local tax to one based on the number of persons within the household. The losers in the system were generally seen to be those on the lowest incomes living in small terraced houses (Figure 7.3). This study also showed the benefits of linking proprietary GIS with a standard spreadsheet package.

It is evident that useful progress *can* be made with GIS in relation to the mapping of deprivation. However, there are more direct indicators of deprivation which are currently missing in terms of their availability at the small-area level. The most obvious of these in the UK are income and GDP, whilst the missing information in the US is social class. These are available at regional scales (from publications such as *Regional Trends*, and *New Earnings Survey*), and Birkin and Clarke (1989) showed how it was possible to build small-area income estimates using microsimulation procedures. The estimates were based on 1981 Census data. Table 7.2 shows the output of recent research to update these figures (shown here for selected wards in relation to larger spatial regions), undertaken as part of a bid to show how areas of Leeds might qualify for EC-assisted status. The figures show the wide variations in income and state dependencies within wards of the city and how these relate to larger geographic regions and the EC as a whole.

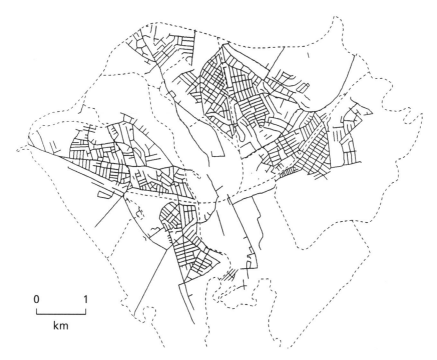

Figure 7.3 Households deemed losers after property tax changes
(*Source*: Martin et al 1992)

The work on income generation also leads to interesting research on benefits and taxation in general. Since income has now been coupled to household structure and occupation in the models, it is possible to apply direct sets of rules to generate state benefits and taxation for each household in the simulated database (see again Table 7.2).

7.5.2 Retail site analysis

One area in which urban models found popularity in the public sector in the 1960s and 1970s was that of retailing (see, for example, Foot 1981). These types of model were used fairly extensively by planners to evaluate the impact of proposed new schemes on existing centres. For example, Foot cites the case of the Haydock proposal in Lancashire, where an ostensibly attractive application for development was rejected for fear of the crippling effect on local centres (cf. Section 5.7).

Now, of course, the political tide has turned somewhat. Thus the potentially damaging effect of the giant new Meadowhall centre on the outskirts of Sheffield in the UK on traders in the centre of Rotherham and Sheffield has not been allowed to stand in its way. Indeed, local planning authorities

Table 7.2 Incomes 1990

Inner-city wards of Leeds	Gross income	Income per household	Income/capita
1 Armley	2 298.1	268.6	104.2
2 Burmantofts	1 998.4	233.3	91.6
3 Chapel Allerton	2 425.5	294.2	107.4
4 City and Holbeck	1 715.7	228.8	96.7
5 Harehills	2 143.6	241.0	87.9
6 Headingley	1 992.3	308.2	135.8
7 Hunslet	1 548.7	244.7	96.2
8 Middleton	2 002.7	262.2	94.2
9 Richmond Hill	1 889.5	239.8	90.1
10 Seacroft	1 931.6	241.0	89.5
11 University	1 618.4	241.7	108.2
12 Kirkstall	2 137.3	261.7	134.5
13 Inner city	23 701.4	254.4	103.0
Other regions:			
1 West Yorkshire	239.6	315.7	117.9
2 Yorks and Hum	554,9	315,0	117,2
3 UK	6 863.6	349.0	128.9
4 EC	38 997.8	338.6	120.5
1 Manchester	179.2	328.3	122.3
2 Liverpool	115.3	321.6	113.3
3 Doncaster	32.0	336.1	121.0
4 Sheffield	70.8	309.5	118.5
5 Newcastle	83.6	286.8	109.3
6 Birmingham	178.8	331.9	118.5
7 Bradford	54.5	313.6	114.4
8 Calderdale	23.1	318.5	121.5
9 Huddersfield	27.5	316.6	119.1
10 Leeds	79.3	305.9	116.8
11 Wakefield	33.1	339.9	123.4
12 Hull	50.0	316.9	116.6
13 Nottingham	92.9	340.0	127.1
14 Glasgow	138.6	312.6	111.0

Note: Income is measured in £000s for inner-city wards, and in £millions for the other regions.

seem to be falling over themselves to make sure that they can get the out-of-town centres, with their promise of new jobs and investment in the local economy: witness the competition to Sheffield's Meadowhall from nearby Parkgate in Rotherham. A potential application area is clearly to discriminate between out-of-town sites. For the developer, this may mean trying to choose between competing sites within different local authority areas – say, Meadowhall versus Parkgate. For the local authority, it may mean choosing between competing schemes within its own boundaries. For example, a private sector report (Thorpe and Partners 1988) showed that at the time of writing there were proposals pending for 3.4 million square feet of new retail shopping in the city of Leeds – nearly one-third of the existing floor space in the city! Not all of these new schemes could be granted planning permission. But which should be supported, on the grounds of efficiency, or perhaps even equity?

We saw in Chapter 5 that the private sector is increasingly interested in GIS for retail store impact assessment. For them, the process is one of defining catchment areas for potential new store locations and allocating the retail expenditures in the catchment area between existing retailers and their new store (cf. Reynolds 1991). So far, there is very little evidence of UK local authorities adopting similar practices in terms of impact assessment. Yet, on the face of it, GIS would appear to offer considerable promise in the field of store assessment research for the public sector. First, it could calculate simple provision ratios to identify areas over- or under-provided for in terms of a range of different retailing activities. Currently this procedure is normally compiled by hand. If gaps in provision have been identified, the GIS may help to explore the underlying geodemographics of the area to ascertain which retailers might be most likely to fill such gaps (an analysis many retailers themselves currently favour – see Chapter 3). Finally, GIS might be used to work out the likely catchment areas of potential new stores and the most likely scenarios regarding impact assessment.

Following the argument of Chapter 3, we would advocate that the impact assessment work would benefit from the integration of models with GIS. Superficially, GIS technology is well suited to turnover estimation for new retail centres. The approach has been demonstrated by Reynolds (1991) for Meadowhall near Sheffield, and by Howard and Davies (1990) for the Metro Centre at Gateshead. The catchment area of the new centre is split into drive time bands, usually of 15 minutes. Within each drive time band a population profile is constructed, using a geodemographic discriminator such as ACORN. For each geodemographic group, an average level of retail expenditure is applied to generate total demand within the drive time band. The penetrations within each drive time band are then estimated, and these penetrations multiplied by the demand to generate the potential revenue within each drive time band. The potential revenue is then summed across each drive time band to produce a total revenue for the centre. A more detailed exposition and critique of this method is provided by Birkin (1994) – see also Chapter 5.

This type of methodology has an appealing simplicity, but it is worth repeating the difficulties. Although one obvious weakness of the approach is that it does not provide any basis on which to appraise the impact of the new centre on existing centres, this was not a major source of concern in the political climate of the UK in the late 1980s (in marked contrast to the Haydock scheme of the late 1960s discussed above, which was rejected because of concern about such impacts – see Foot 1981). However, there are other serious problems with the methodology. One is that it is simply not realistic to assume a uniform rate of penetration within each drive time band. For example, studies of the Metro Centre have shown a marked variation in penetrations between the northern and southern parts of the catchment (to the north, the presence of the city of Newcastle as a competing centre imposes a marked depression on penetrations – Davies and Howard 1989). A second, and related, difficulty is how to estimate the penetration levels within each drive time band. Davies and Howard used an analogue to try and forecast Meadowhall's turnover on the basis of experience of the Metro Centre, but their estimates varied wildly according to whether the analogue was based on the impact of the Metro Centre to the north of the River Tyne on the one hand (giving a trading estimate for Meadowhall of £274 million), or to the south of the river on the other (giving an estimate of £480 million – Davies and Howard 1989: Table 6).

The main source of problems with the catchment area method is in fact a failure to consider impacts, because the turnover potential of any new centre will be dependent on the switching of customer expenditure away from existing centres to the new centres. In order to quantify this behaviour, an accounting framework is needed which combines elements of consumer demand, retail provision, and accessibility. Once again, the type of framework defined in Chapter 5 is the most appropriate way to achieve this. The method has been implemented in a (post-opening) study of the Thurrock Lakeside centre, Essex, UK. This was opened on 23 October 1991. The difficulty with adopting some kind of catchment-cum-analogue methodology (as outlined above) can be illustrated in this case with reference to Table 7.3. This table shows that the Lakeside Centre is broadly comparable to both Meadowhall and the Metro Centre in its size and composition, while having more than six times the catchment population of either centre within 30 minutes' drive time. These data imply that we would expect Thurrock to out-perform the other two centres quite comfortably, and yet the evidence is that Thurrock has in fact been the least successful of the three (although unfortunately this evidence comes in the form of confidential sales data for individual GMAP clients, which cannot be disclosed here).

Table 7.3 Comparison of out-of-town centres

	Metro	Meadowhall	Thurrock
Gross retail and service space	1.4m	1.1m	1.3m
Car parking	10 000+	12 500+	9000+
Estimated population in 30 minutes' drive time	542k	547k	3.6m

In order to quantify the impact of the Lakeside Centre, we have used a spatial interaction model which identifies retail floor space in each of 1743 centres in the UK, and estimates customer demand in each of 2625 postal districts. Flows of expenditure from districts to centres are mediated by the drive time from each postal district to each centre (so this is a much more detailed approach than grouping areas together into drive time bands – once again, see Chapter 5 for a fuller description of the methods). The model was already in existence in a calibrated form before the opening of the Lakeside Centre. It was therefore possible to simulate the impact of the new centre by creating a new destination within the model with the appropriate characteristics. We used customer samples drawn from Thurrock over the first five months after the new centre was opened, and turnover data from retailers within the centres to estimate the turnover of the Lakeside Centre and its penetration within surrounding postal districts.

The variation in market penetration against drive time, both observed and modelled, is shown in Figure 7.4. The model clearly has a tendency to smooth the fluctuations within the observed data, and of course it is always in the nature of a model to simplify real patterns to an extent. Equally, it is in the nature of any survey sample to produce exaggerated variations which might tend to be smoothed by more complete data. We would argue (strongly) that Figure 7.4 is a much better representation of variations in market penetration than could be achieved by an approach which lumps together customers within drive time bands. This is also shown in Figure 7.5, where a scatterplot of observed against predicted penetrations is shown for individual postal districts.

As with the catchment analogue method, the most difficult issue is how to estimate the total turnover for the new centre. Our experience is that new out-of-town centres are more attractive, foot for foot, than city centres with equivalent retail provision. Factors such as the amenity of parking and existence of complementary attractions (such as cinemas and children's play areas) and the generally attractive layout of the new centres must all play a part. In effect, we need to develop an analogue within our own spatial interaction model which relates the attractiveness of new centres to each other, and to existing centres. Our calibrations have shown that each square foot of floor space in a new centre is worth between 30 per cent (Metro Centre) and 60 per cent (Meadowhall) more in an out-of-town retail centre than in a typical city centre. The early evidence was that Thurrock Lakeside had an intermediate performance, at around the 50 per-cent mark.

Once the overall attractiveness of the centre has been estimated, we can use the model to estimate deflection patterns from neighbouring centres. In this case, we were able to attempt to validate the model using time series data which were available for stores in a number of surrounding centres. Some results are shown in Tables 7.4 and 7.5. Again, this is a difficult task

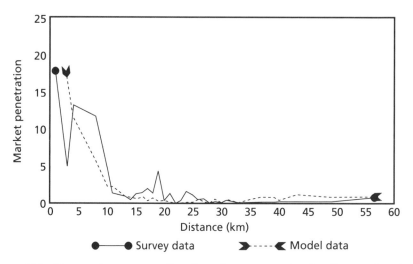

Figure 7.4 Observed versus predicted market penetration by distance band

because fluctuations in the revenues of these stores will be driven by factors other than the new centre opening. For example, the opening of the new centre appears to generate an increase in the observed revenue in the other centres for one product line, and this is clearly counter-intuitive; while stores in both Palmers Green and Brentwood actually gain revenue, which again cannot be directly ascribed to the new centre opening. Nevertheless, the general pattern is an encouraging one in the sense that we are able to identify accurately both the products for which the deflections are most significant, and also the stores which suffer the most impact.

In conclusion, it should be stressed there are difficulties, particularly in assessing the likely attractiveness of a new centre, and hence the probable penetration of the surrounding catchment. Nevertheless, spatial interaction modelling provides a comprehensive and appropriate framework for comparative analysis of new retail centres, and a means for assessing their impact on the surrounding area in terms of both competing stores and catchment penetration. Doubtless these models could be considerably improved through further research.

7.5.3 Labour-market analysis

7.5.3.1 Local labour-market research

The most obvious advantage of GIS is the ability to store, retrieve, and map a large volume of data relating to a system of interest. In the study of local labour markets there are a number of examples of large-scale information

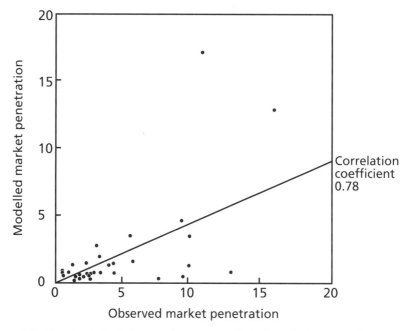

Figure 7.5 Scatterplot of observed versus predicted market penetration

systems which combine Census data with other information on jobs, industries, and individual plants. Worrall and Rao (1991) describe the Telford
information system used to monitor job losses and gains and the changing
patterns of journey to work, which have increased in length following the
construction of the M54 motorway. Their GIS is used to perform spatial
components of change analysis, thus identifying the locations where job
losses and gains have taken place and monitoring the catchment areas of
large employers, industrial estates, or enterprise zones. Hart et al (1991) add
skills audits (cf. Haughton and Peck 1989) to their information system in
Belfast to help identify catchment areas for a workplace, on the basis
of likely limits of travel to work for different skills groups. By using the

Table 7.4 Observed versus predicted deflections by product type

	Predicted (£000s)	Observed (£000s)
News	−27.4	72.9
Books	−275.7	−295.9
Stationery	−225.6	−266.0
Records and video	-182.4	-98.6
Cards	−35.7	-85.4

Table 7.5 Observed versus predicted deflections by store and by product type

Name	Product 1		Product 2	
	Observed	Predicted	Observed	Predicted
Romford	17.5	−10.6	−129.3	−92.3
Basildon	−11.1	−6.1	−78.7	−63.8
East Ham	7.0	−1.1	−37.1	−17.5
Chelmsford	−25.8	−3.1	−30.0	−31.7
Stratford	14.0	−0.6	−22.0	−10.0
Southend	32.9	−2.9	−15.6	−26.3
Palmers Green	0.1	−0.1	8.4	−1.2
Brentwood	38.3	−2.9	8.4	−32.9

Note: All figures are in £000s.

standard polygon overlay facilities within a GIS, the planner could combine unemployment data with the findings of skills audits to work out the optimal job mix for a particular region.

As identified above, new developments in an area are seen as attractive for the purposes of economic development; that is, new jobs are provided. It is then necessary to ask what the impact of these new jobs will be on a local economy: where will the employees live, and what additional demands will they make on the existing transport network? Furthermore, if a site is to be redeveloped then one kind of control which the authorities can still exercise is over the type of new development: should it be office space, or manufacturing, or a retail park? Again, the answers to these kinds of questions can be more fully answered by linking GIS with model-based analysis.

Take Bradford city council, for example, which is currently negotiating the redevelopment of the site of its transport museum at Low Moor. This is a relatively deprived area of Bradford, with high levels of unemployment and poor-quality service provision. However, to redevelop the land as something like offices might do little for the local unemployed population: employees would simply be sucked in from the relatively prosperous suburbs. The construction of a new manufacturing establishment – a car assembly plant, perhaps – would be far more beneficial, if only it could be attracted. It is thus vital to try and estimate the likely catchment area of any new development in order to predict the likely impacts on *local* unemployment. Birkin et al (1990) demonstrated this approach for 100 new hypothetical jobs in Leeds, whilst Figure 7.6 shows the predicted catchment area for a new science park (and hence high-technology jobs) at Weetwood, north Leeds.

Clearly there may also be implications for adjustments to the transport network at the local level (planning which junctions need to be improved in Low Moor itself) and for traffic flows across the metropolitan area (can existing links take the extra burden?). However, such judgements would

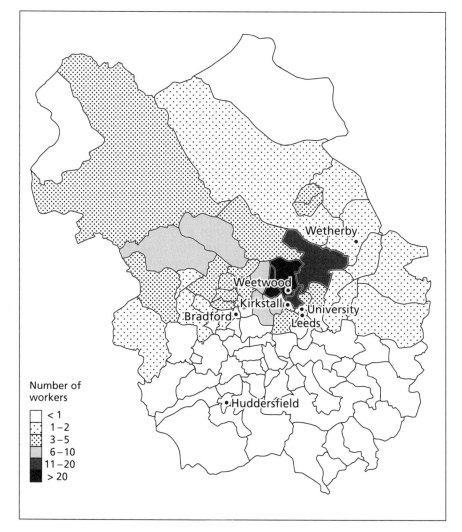

Figure 7.6 Predicted workforce catchment area for a new science park
(*Source*: Birkin et al 1994)

usually be taken on the basis of crude and inadequate information – simply the number of new jobs created, and a pro rata extrapolation of existing journey-to-work flows.

7.5.3.2 Location and operation of Training and Enterprise Councils

In December 1988 plans were announced in the UK for the creation of locally based Training and Enterprise Councils (TECs) to support and promote the provision of training and business de᷈elopment. From April 1990

TECs were to be administered and operated by new, local, employer-led bodies, consisting mainly of local private sector employers with additional support from local authorities and the voluntary sector. Their brief was to work towards the implementation of a set of national priorities on training, but also to take account of individual peculiarities within local labour markets. For this latter reason, a key emphasis in the literature on TECs (see Evans 1992, Boddy 1992, Green and Owen 1992) has been the need for them to respond more flexibly to local needs and circumstances. This has made the study of the geography of local labour markets, and their relative performance, of direct relevance to the understanding of the performance of TECs. This is because training, like most other service activities, is largely delivered from fixed-point locations, and these are more likely to be the result of historical legacies than of an up-to-date geographical appraisal. In addition, demand for training is unevenly spread across any city or region. This arises out of the uneven residential distribution of the various categories of unemployed or low-skilled workers, as well as the uneven distribution of company types which themselves demand training services (often more specialist). Hence, within any TEC boundary, there will be variations in both training needs and the patterns of supply. As a consequence, we believe that there is an important and necessary obligation for TECs to develop integrated information concerning the level of variation of these key variables. In addition, there is a requirement for effective planning tools to assist in the most efficient and productive use of resources.

As far as SDSS or IGIS are concerned, the requirement is to bring together all the information available in an individual TEC area (for a review see Green and Owen 1992), and to allow TEC managers not only to analyse the existing local labour and training market but to make forecasts of the likely impacts of changing that training market in some way. The SDSS should include:

- a powerful database-handling system containing data on potential 'client groups' (unemployed, etc.), employers, and training providers specified at a very fine geographical scale. The idea is to bring together the variety of information from various 'field systems' on to a single, easy-to-use PC system which contains attractive mapping and graphics facilities;
- a predictive modelling capability that allows the impacts of changes in a number of variables to be assessed, particularly those relating to changes in training provision;
- the development of a set of performance indicators that allows current effectiveness of training provision to be assessed, as well as the identification of gaps between training need and supply. These indicators also help evaluate the impact of proposed plans.

The main information components of the GIS can be summarized under a number of headings. First is 'economic activity', including population and household data (the workforce). One clear indicator of training needs is the level of residential unemployment. This will be especially important when trying to interpret 'demand' for training by small geographic area. The variation in unemployment rates across the city is marked and vitally important to understand. Figure 7.7 illustrates the spatial concentration of unemployment for, as an example, the long-term unemployed. Maps such as this clearly bring home the importance of targeting training information and monitoring the success of (re)training by monitoring how such social inequalities can be reduced. This is also a crucial part of the modelling exercise described shortly. Additional workforce data sources could be added in this category as TECs build up their information base. Skills audits of the local population are one obvious area where much work is currently being undertaken (see Green and Owen 1992, Haughton and Peck 1988).

The second component of the SDSS or IGIS includes information on the workplace and its dynamics. It is essential for training providers to have a deep understanding of how the local labour market is changing across the city as a whole. Through a combination of the numerous Censuses of Employment, a detailed picture of employment change by postal sector exists. This can be linked to establishment databases in order to pick out important changes at the level of the individual firm. This is important when considering likely training needs for persons within existing compa-

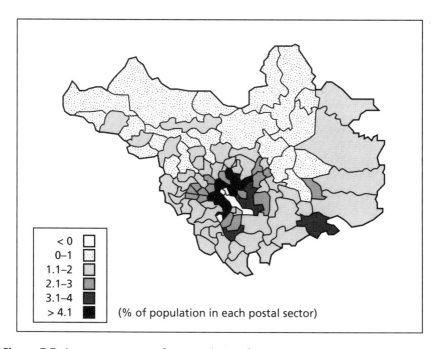

< 0	
0–1	
1.1–2	
2.1–3	
3.1–4	
> 4.1	(% of population in each postal sector)

Figure 7.7 Long-term unemployment in Leeds

nies. It is also useful for identifying *new* firms in the city and asking questions about the types of job these firms are creating, with all that implies for the provision of training.

In addition, it is important to know how dependent local communities are on certain key players in the labour market. If only a few key firms employ the bulk of local workers, it is necessary to keep tracks on those industries in general and the firms in particular. There is a powerful argument, therefore, for using small-area journey-to-work statistics, in conjunction with the Census of Employment and establishment data sets, to estimate the local population dependent on key firms. TECs could then work towards having some sort of contingency plan if large-scale redundancies resulted from key firm closures or cutbacks. At the very least they would be able to estimate quickly the type of person (in terms of skill levels) affected and the other existing local opportunities available.

The third component of the information base deals with training providers themselves. This makes up the supply side of the training network. It is important to see how well a city is served for different types of training and who is actually providing that training (private or public sector). If training of a certain type is not available in the city, how far do people have to travel (daily or on a one-off basis) to obtain that service? Conversely, how far are people travelling into the city to obtain services not available elsewhere? If a training course is not available locally (and many local companies are having to send employees away from the region), can TECs help stimulate such courses to be offered locally (and perhaps help bring more people into the region to be trained)? It is necessary, therefore, for a TEC to have a measure of flows into its territory and outflows from its territory. It is also useful to compare a TEC region with others across the country: what skill shortages are apparent in neighbouring TECs? In this way, it may be possible to think about demand-led training: that is, can a TEC encourage more courses on skills needed in the wider economy rather than just the local labour market?

In addition to displaying spatial information on the training and labour market of an area, we have also recognized the importance of offering the ability to examine the impacts of possible new developments in the area through a suite of models and performance indicators. A suitable model can be written as:

$$S_{ij} = A_i \times O_i \times W_j \times f(c_{ij}) \tag{7.1}$$

where:

S_{ij} is the flow of people from residence or workplace i to training centre j;

O_i is the potential demand in area or workplace i;

W_j is the number of courses on offer at centre j;

c_{ij} is a measure of the cost of travel or distance between i and j;

A_i is a balancing factor which takes account of the competition and ensures that all demand is allocated to centres in the region. Formally it is written as:

$$A_i = 1/\sum_j W_j \times f(c_{ij}) \tag{7.2}$$

The performance indicators can again be seen as residential indicators, which help to show the variations in access to training for different target groups in the city, and/or facility indicators, which relate to the performance of the training providers. Details of these types of indicator are provided in the appendix.

The model can be run in two different modes of operation. The first is called the *baseline* and relates to the existing patterns of demand and supply. For this present-day structure we can see how well different areas and different sorts of person (or target group) are served by existing facilities, and where important gaps remain. The second mode of operation is the *new scenario* model, which allows the user to input the addition of new courses in certain locations, or simply increase the range of courses presently on offer. The impact of these changes can be seen through the new set of performance indicators, which will indicate the changes in accessibility and provision ratios caused by these additional facilities. In this way, a more proactive management structure can be created.

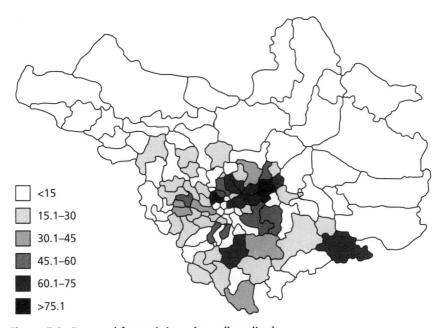

Figure 7.8 Demand for training places (baseline)

Legend:
☐ <15
▨ 15.1–30
▨ 30.1–45
▨ 45.1–60
■ 60.1–75
■ >75.1

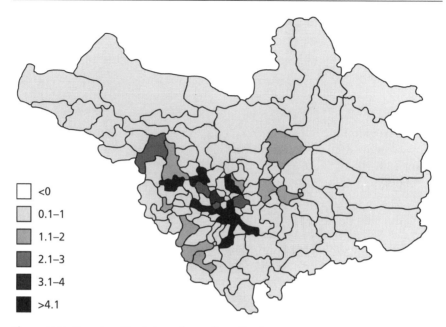

☐	<0
☐	0.1–1
☐	1.1–2
■	2.1–3
■	3.1–4
■	>4.1

Figure 7.9 Supply of training places (baseline)

At this stage, an example using information built up for Leeds is useful. Suppose we are interested in the pattern of demand for training by young unemployed persons and the provision of courses for that group of the population. Having run the base model, the basic demand pattern and the location of available courses can be mapped (Figures 7.8 and 7.9 respectively). From the basic suite of performance indicators we can look at the nature of the relationship between demand and supply. Figure 7.10 shows a very simple indicator (to fix ideas): the coincidence of significant demand levels and long distances to training providers (residential indicator 3 in the appendix). There are clearly major gaps in provision and a number of obvious problem areas: south-east Leeds, north-east Leeds, and north Leeds. In terms of actual numbers of unemployed, the south-east Leeds area is undoubtedly the worst (compare Figures 7.8 and 7.10).

This is one use of the models – to find areas poorly served by existing levels of training provision. After this, the likely consequences of changing the level of provision in some way can be explored. Figure 7.10 highlighted the problem in south-east Leeds. The new-scenario model allows us to look at the issue of adding new courses in this area. Figure 7.11 shows the new supply pattern when we add five new courses in this part of south-east Leeds (compare to Figure 7.9). Note that the demand for training is still the same – only the supply level is being changed. When the model has been run for this new situation, the performance indicators can be mapped

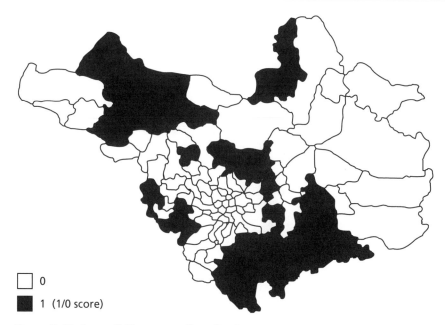

☐ 0

■ 1 (1/0 score)

Figure 7.10 Accessibility scores (baseline)

again and compared to the previous run. Figure 7.12 maps the new set of accessibility scores.

Clearly the impact on (in)accessibility is large in this area of high young-person unemployment (compare Figure 7.11). However, there is also a very steep distance-decay effect away from the new supply point: the model shows that this new level of supply will not make great inroads into meeting training requirements in this area of relatively heavy demand. This is due to both the large numbers currently unemployed and the fact that low-skilled persons do not (or cannot) travel long distances to take up employment or training. That said, it is clearly preferable to simply adding more courses in existing centres many miles from this part of Leeds.

In the example above, only one simple indicator has been used, but it is hoped that the policy implications are clear. The system can be used to evaluate the current relationships between training supply and potential demand, and thus to identify both target areas where existing levels of training provision are least adequate, and target groups of clients for whom improved levels of training provision are most urgently required. From this base, new scenarios regarding the addition of new facilities (or indeed closures or relocations) can be quickly and effectively judged.

The model outputs can also be used to examine the performance of individual training providers in more detail. So, for each provider the analyst could:

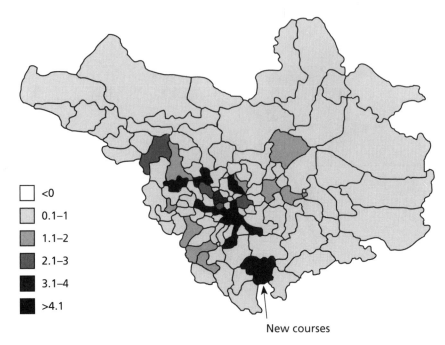

Legend:

☐ <0
☐ 0.1–1
▨ 1.1–2
▦ 2.1–3
■ 3.1–4
■ >4.1

New courses

Figure 7.11 Supply of training places (new scenario)

1 estimate catchment area;
2 provide summary indicators for that catchment area: that is, numbers of residents in target groups in relation to the number (and type) of courses on offer, compared with national averages.

This would target performance measures down to the very local level and provide interesting statistics for the individual providers of training. In addition, there is another sort of modelling exercise to be attempted here: the ability to change the make-up of courses to maximize local participation rates. This could be used in the spirit of a planning game exercise, where the user could experiment with a wide range of course options given the types of resident (by target groups) in the catchment area (including how the catchment area itself may vary if a different range of courses is offered).

7.5.3.3 Employment office location

The changing information requirements of urban planning in the 1990s may be facilitated by technological and analytical advances, but an even more important consideration is the changing political and institutional framework. In the UK, one of the most important trends in the 1980s and early 1990s has been an increased specialization, or perhaps fragmentation

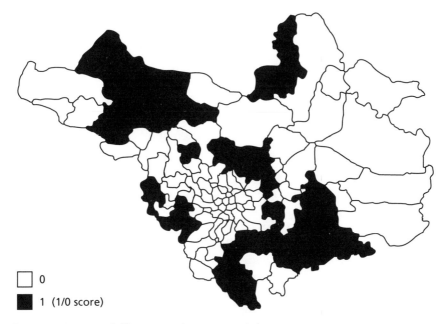

□ 0

■ 1 (1/0 score)

Figure 7.12 Accessibility scores (new scenario)

(depending on one's point of view), in the provision of the services of local government. A good example of this trend is the creation of the Employment Service (ES). This is a national agency which is hierarchically structured into regional and sub-regional divisions, with four key areas of responsibility:

1 payment of benefits to appropriately qualified ('unemployed') claimants;
2 provision of (re)training to job-seekers;
3 provision of training to employees within the workplace;
4 matching of job vacancies with suitably qualified job-seekers.

Historically, these services have been provided through two different networks:

1 Unemployment Benefit Offices (UBOs), through which benefits are paid;
2 Job Centres (JCs), which are foci for the training and placement activities.

A primary agenda for the ES is how to increase the efficiency of this network of offices, perhaps through structural changes such as the combination of training, placement, and benefit functions within single integrated offices, or perhaps simply through performance monitoring to ensure best practice within the offices. It is within this context that GMAP was commissioned by the ES to provide a prototype decision support system for the Bradford area. The following objectives were identified:

- to measure the performance of local offices, both currently (between offices) and over time;
- to assist local offices in better targeting of their services, given the registration profile of that office – that is, to help offices to identify the most effective combination of training programmes taking into account the skills base of the existing pool of claimants, and the needs of local employers;
- to identify areas of under- or over-provision of offices;
- to determine the optimal locations of new offices, and predict the levels of business which such offices would generate.

In order to meet these objectives, a system with two components, a GIS and a geographical modelling system, was produced. Within the GIS, information is presented at three spatial levels: postal sectors, of which there are 184 in the Bradford region, ES offices (19), or 'towns' (11 as defined by the ES). The data presented include a variety of information on employers, employees, claimants, placings, programmes, and performance. As usual, the key features of this stage of the process are that it involves the combination of different databases from the ES together with data from the Census of Population and the Census of Employment; the construction of appropriate performance indicators from these data; and the presentation of the data in user-friendly maps, tables and graphical formats.

The modelling system itself contains three further options:

1 a *scenario model*, which examines the effect of altering the current network of offices in some way. The model also allows the allocation of claimants to offices (either existing or revised) to be optimized. The scenario model uses known patterns of UBO usage by claimants ('Livingstone data') to calibrate a model of spatial interaction between claimants and offices. The model options include facilities for closing, opening, or relocating offices, as well as altering the size or staffing levels of offices. An example scenario involving the closure of an office in the town of Batley is shown in Table 7.6;
2 a *reallocation model*, which looks at the effect upon the performance indicators of changing the allocation of a particular group of claimants from one office to another;
3 a *location model*, which finds the optimal number and location of offices given the current distribution of claimants and other performance criteria, such as the ideal size for offices and the maximum distance which it is reasonable to require claimants to travel.

While the current applications of the model concentrate on benefit claims, we hope to incorporate other features, such as the location of training programmes or the effects of a changing unemployment pattern on network

Table 7.6 Impacts of closing Batley employment office (*Source*: GMAP 1994)

Town name	Baseline total claimants	Scenario total claimants	Deflection rates
Batley	1 802	0	−1 802
Bradford	17 756	17 756	0
Brighouse	1 145	1 145	0
Dewsbury	4 167	5 653	1 486
Halifax	6 671	6 671	0
Huddersfield	8 760	8 760	0
Keighley	3 608	3 608	0
Shipley	2 303	2 303	0
Skipton	1 113	1 113	0
Spen Valley	2 003	2 319	316
Todmorden	1 161	1 161	0
Total	50 489	50 489	0
Mean	4 590	4 590	0

performance, in future developments of the system. Linkages to the labour market and training applications discussed in Sections 7.5.3.1 and 7.5.3.2 above are obvious.

7.5.4 Population forecasting

In order to be able to plan effectively for the development of the housing stock, provision of services, and changing infrastructure requirements, planning agencies require small-area population forecasts. Typically, local authorities will need at the very least ward-based population forecasts, these being one of the base units of British Census and electoral geographies (each ward contains roughly 10 000 households, and a typical urban district would be broken into between 20 and 30 wards).

However, existing approaches to population forecasting are relatively crude. OPCS produces district-level forecasts. Many local authorities will simply break these down into wards in proportion to existing or last-known (Census) distributions, with some allowance for the existing demographic mix. (There are exceptions to this – Woodhead (1985) reports that certain authorities do use more sophisticated cohort-survival methods – but these are few and far between.)

This type of procedure ignores a variety of data sources which can be used to help generate better estimates. Thus OPCS also produce annual vital statistics on births and deaths by ward, which can be used to constrain and modify district-based fertility and mortality projections for existing wards. However, the key process for demographic development within the city is both inter-urban and intra-urban migration. Full inter-ward migration matrices are again only available from the Census, and even then they need to be broken down by age and sex. Intra-urban migration can, though, be linked to the changing availability of housing stock through local authority house-building and demolition statistics; and can be linked to changing district-level migration propensities through the National Health Service Central Register (NHSCR – for instance, Boden et al 1992). Finally, as well as forecasting, it is also typically important to consider the question of updating from a previous Census. Beginning with a base population at a Census year, other data, specifically electoral registration statistics, can be used (with care!) to help to provide an accurate, up-to-date picture of small-area population distributions.

The mechanics of the population-modelling procedures need not concern us here (more detail can be found, as desired, in Rees and Rees 1991). The central point to be made is that relatively good forecasts can be generated through the *non-trivial* combination of a variety of data sources. For example, migration may be best represented through some kind of spatial interaction model (cf. Stillwell 1994). The models also provide a 'what if?' capability to analyse the impacts of changing political strategies or demographic behaviour. Hence it becomes relatively easy to forecast the effects of changing housing policy, or alternative assumptions about future patterns of fertility.

These techniques have been applied recently in the Swansea Information System (SWIS – see Rees and Rees 1991) and a variety of private sector clients. The Swansea system is based on Census ward geography. The private sector is increasingly interested in postal district level (since this is the usual spatial resolution of its customer data). In technical terms, this poses new challenges. The research task is to disaggregate known local authority estimates to the postal district level. Future births and deaths at the postal district level can be forecast using the usual cohort-survival methods, since Census data are now available at this level of spatial resolution. Migration can be estimated from a combination of Census data and the NHSCR data mentioned above. Alternatively, a cruder method would be simply to trend known migration statistics forward (trending forward the 1981–91 known rates) whilst taking into account known house construction or demolition. Figures 7.13 and 7.14 (Plates 7.1 and 7.2) show the results of such small-area forecasts for Scotland and central England respectively. The depopulation of the metropolitan areas is evident in forecasts to the turn of the century. These forecasts can then be entered into the normal spatial models of consumer markets to ensure that an investment made at the present time will still be in an advantageous location in 10 or 15 years' time.

7.6 CONCLUDING COMMENTS

Section 7.5 has argued that it is possible to apply urban models to a variety of planning problems. However, there is a stronger argument, which is that this type of comprehensive application is not only possible but absolutely necessary, because of the interdependence between these systems of interest. For example, it is interesting to prepare demographic forecasts in isolation, but these turn out to depend crucially on small-area migration flows. It is surely not possible to understand these migration patterns without reference to the changing availability of jobs within the city, and the various types of provision of services and amenities. And again, it may be interesting to compare the relative merits of derelict sites as potential shopping-centre locations, but a key issue is what type of land use should be approved for a particular site. This needs to be evaluated with respect to the impact within different sectors of the economy – retailing, services, manufacturing, leisure, transportation, and so on.

At one level, this may sound like a call for the return of the old integrated (but failed!) land-use and transportation models of the 1960s. But why not? The spirit of our argument is that the technical and political climate is ripe for such a revival and that this type of applicability cannot be provided by existing GIS.

CHAPTER 8

Water planning and GIS

8.1 INTRODUCTION

> Water is an important resource being used in a variety of ways at many different levels which produces social, spatial and organizational problems. Given also its potential as a source of hazard, the result is a management task of remarkable complexity.
>
> (McDonald and Kay 1987:1)

Figure 8.1 shows a schematic representation of the main components of the water supply system. It shows the sequence of events from the capture of water in reservoirs through the demand cycle to the final removal of waste. It also shows that there are a significant number of interactions in the water supply system. A number of these will be explored below.

In most countries, the provision of potable water and the disposal of sewage products are undertaken by utility companies that are either publicly or privately owned. For operational reasons, they tend to enjoy a monopoly of service provision over a region of a country. The planning task of a typical water company can be viewed as the integration of three different areas:

1. *hydrology* – understanding the mechanisms by which the raw material is generated and modified;
2. *engineering* – the control and transformation of the raw material so as to meet demand by consumers;
3. *economics* – the cost associated with maintaining a supply to the consumer of acceptable quality, and the subsequent pricing of that supply to those consumers.

Geography becomes important in all three areas as the supply and quality of water varies across a region; the engineering aspects control the supply/demand interaction across space, and the costs of all these operations also vary across space.

The aim of this chapter is to review the use of GIS in the water industry and to examine the integration of economic, engineering, and hydrological modelling for effective water resource planning, particularly in the UK. It is useful to begin by briefly recounting the key actors in the water industry and the roles they each play. The water companies in the UK, like many elsewhere in the world, are responsible for water provision across large

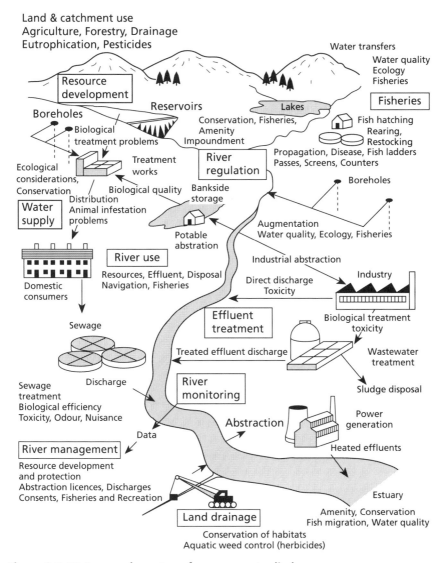

Land & catchment use
Agriculture, Forestry, Drainage
Eutrophication, Pesticides

Water transfers
Water quality
Ecology
Fisheries

Resource
development

Fisheries

Boreholes

Reservoirs   Lakes

Biological
treatment problems

Conservation, Fisheries,
Amenity
Impoundment

Fish hatching
Rearing,
Restocking
Propagation, Disease, Fish ladders
Passes, Screens, Counters

Treatment
works

River
regulation

Ecological
considerations,
Conservation

Boreholes

Water
supply

Distribution
Animal infestation
problems

Biological quality

Bankside
storage

Augmentation
Water quality, Ecology, Fisheries

Potable
abstraction

Industrial abstraction

River use

Industry

Resources, Effluent, Disposal
Navigation, Fisheries

Direct discharge
Toxicity

Domestic
consumers

Sewage

Effluent
treatment

Biological treatment
toxicity

Treated effluent discharge

Wastewater
treatment

Sewage
treatment
Biological efficiency
Toxicity, Odour, Nuisance

Discharge

River
monitoring

Sludge disposal

Abstraction

Power
generation

Data

Heated effluents

River management

Resource development
and protection
Abstraction licences, Discharges
Consents, Fisheries and Recreation

Estuary

Land drainage

Amenity, Conservation
Fish migration, Water quality

Conservation of habitats
Aquatic weed control (herbicides)

Figure 8.1 Water supply system: from source to discharge
(*Source*: Newson 1992: xxv)

regions, although there are a number of smaller companies usually responsible for the provision of water services to a single town. These companies
are now in private ownership. For this reason the government has imposed
a number of regulatory agencies to ensure that the consumer does not
suffer through a poorer service (both in quality and through higher prices).
Figure 8.2 shows the structure of the regulators of the private market and
their respective roles.

The National Rivers Authority (NRA) has most direct interest in environmental and pollution aspects of the water industry. Its role is to license the abstraction of water from any source: uplands, boreholes, or rivers. In addition, it must set consents or standards for the quality of returned waters to the environment. In sum, its role is to 'police' UK rivers, license abstractions, and bring polluters to justice.

The second major regulator is OFWAT (Office of Water Services), which has a broader remit. It is required to monitor the activities of the water companies to ensure the delivery of an adequate quality of water at a fair price to the consumers, whilst at the same time allowing a reasonable return to investors in the water industry. There is still an effective geographical monopoly on water service provision in the UK despite privatization, and measures of price control are clearly required. The third major regulator is the Drinking Water Inspectorate (DWI), which focuses primarily on water quality. The minimum standards are laid down by the European Commission and the Department of the Environment in the UK. Hence the DWI's role is to monitor the extent to which adherence to standards is maintained.

In addition to these public and private bodies, there are a number of external agencies which act as research consultants. The largest is the Water Research Centre, which employs 600 staff whose remit covers all aspects of

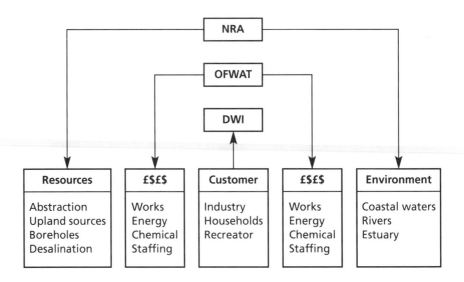

NRA = National Rivers Authority

OFWAT = Office of Water Services

DWI = Drinking Water Inspectorate

Figure 8.2 Structure of UK water regulators since privatization
(*Source*: GMAP 1994)

water planning, from water treatment and distribution to sewage and waste disposal. Kingdom (1991) describes the involvement of the Water Research Centre with assessments of the role and potential of GIS in the water industry. The rest of the chapter reviews the role of GIS within all these branches of the industry.

8.2 GIS AND THE WATER INDUSTRY: A REVIEW

The complexity alluded to by McDonald and Kay above has traditionally provided a rationale for investigating the use of spatial modelling procedures in water resource planning. Typically, the complexities are investigated using simulation models (either time-sequenced or event-sequenced) and optimization and regression techniques (see McDonald and Kay 1987, Dzurik 1990, Newson 1992). Examples in the UK include FRQSIM, a package which simulates the frequency of storms, and ONDA, a model to estimate water flows through channels (see Thomasson 1989). In the US the Storm-water Management Model (SWMM) is one of the most frequent tools used for simulating storm conditions and their effects on flood plains and sewer/storm-water drainage systems (Dzurik 1990). Such simulation models make natural partners for GIS, given the latter's capabilities in terms of data visualization. Streit and Wiesmann (1993) report the interest shown in linking GIS to hydrological models at the first international HydroGIS conference held in Vienna in 1993. They report that more than 20 papers were devoted to the tighter integration of models and GIS (see also their own contribution to the integration of these models into GIS through either direct or indirect coupling).

In this way, such environmental applications of GIS seem to be ahead of many of their counterparts in more mainstream human geography applications, where the coupling of more sophisticated analytical functions has still to take place (cf. Chapter 3). It is useful to review this progress under a number of headings.

8.2.1 Network analysis

The experience of GIS for many water companies has initially come through the digitization of channel and pipeline networks. This is as true for the water industry as it has been for all the other utilities (see Rector 1993). Indeed, the transformation of basic infrastructure information from paper to digital records provides the GIS industry with perhaps its largest market world-wide. Elkins (1990) describes Southern Water's digital mapping exercise, which involves the capture of all asset records, including 11 500 km of water mains and 24 600 km of sewers within an area made up of 16 000 OS maps. Once constructed, such GIS can assist in all aspects of network planning. Kingdom (1991) describes their use for 'rehabilitation planning': the consideration of structural and hydraulic performance of pipe systems to facilitate plan design for maintenance and repair.

At a more strategic level, there are a number of projects which aim to link GIS with spatial models in order for river networks to be better understood in relation to the effects of climatic change and to water quality. Section 8.2.2 returns to the latter issue. The Institute of Hydrology in the UK has digitized the entire UK river network, which itself has been used to form the base of a digital terrain model of the UK. The combined Water Information System is used not only for the study of flow dynamics but also for river-quality mapping, derivation of catchment characteristics, and low-flow estimations (Finch 1992). Screen graphics facilities allow the user to roam up and down a river network to examine related environmental data.

Finch (1992: 19) also describes the attempts of the Institute of Hydrology to integrate GIS with hydrological models. He describes the objectives of a new project as follows:

> to provide core models for predicting impacts of climatic change on biogeochemical and ecological systems; to provide models which run for both equilibrium and transitional climates; to couple the models with a GIS to examine the impacts spatially across the UK.

Although such national models are important planning and monitoring tools, a number of the regionally based water authorities have increasingly recognized the need for more local information systems for network planning. This demand has been stimulated by the transformation of water transfer from single source/single destination systems to multiple source/multiple destination systems integrated over entire river basins. This in itself was partly stimulated by the severe droughts in the UK during the summers of 1976 and 1984, which highlighted the need for water to be switched between source points in order to satisfy demand in the most severely affected regions.

The realization that such systems were not yet commercially available led Yorkshire Water PLC to commission the construction of a new simulation tool for water resource planning by GMAP. GRIDNET, the subsequent PC-based resource planning tool, provides a simulation model for a range of operational planning and management activities, including analysis and growth of demand, effect of leakage control, analysis of hydrological change, changes in reservoir operating rules, variations in abstraction licences, and the overall impacts of changes in land use or catchment management policies. In addition, it offers the type of high-quality graphics and data storage/management facilities expected in any GIS.

GRIDNET is a decision-support system that allows the user to specify the characteristics of the various water sources (reservoirs, boreholes, river abstractions) the network, and other components such as treatment works, and run various scenarios over a 40-year historical data set of weekly rainfall. The objective is to operate the grid, shown in Figure 8.3 (Plate 8.1), in a way that attempts to minimize operating costs. As can be seen, there are many possible ways of delivering water from sources to sinks, and the costs of doing so vary enormously. For example, river water abstracted from Elvington has to be pumped to meet demand in west Yorkshire. River water is also more costly to treat than upland reservoir water, and therefore its use is typically minimized.

GRIDNET is made up of separate modules that include:

- a *reservoir simulation model*: this model operates a set of rules that relate to the amount of water that can be abstracted daily from a reservoir given certain conditions relating to the level of water in the reservoir and whether the reservoir is filling or emptying;
- a *pumping station model*: this simulates the operation of each pumping station in the network under different throughput conditions. In particular, the costs of operation are determined under different conditions and electricity tariffs;
- a *treatment works model*: this simulates the operation of a treatment works in terms of the quality of the water that enters, the capacity of the works, and the cost of treatment, especially the cost of chemicals used to bring water up to an acceptable quality standard;
- a *network optimization model*: this component integrates the outputs of the three previously described models and attempts to fund a solution to the problem of allocating water from sources to demand through the use of a heuristic modelling procedure.

The model is run over a 40-year historical data set of weekly rainfall, so that the operating costs of the network in a particular scenario can be assessed under a range of likely operating conditions. In this way the cost of optimal network operation can be derived for:

- an average year (the average of the 40 different solutions);
- the wettest year on record;
- the driest year on record.

In this way, upper and lower bounds can be put on expected costs as well as an average. Some outputs from GRIDNET now illustrate how it can be used in practice.

Table 8.1 shows the opening screen menu. To illustrate its capabilities we will describe one policy simulation based on the problem of planning for water shortages given adverse drought conditions. The sequence begins by choosing the 'control line program' option from the opening menu in Table

Table 8.1 Screen menu in GRIDNET (*Source*: GMAP 1993)

WELCOME TO GRIDNET

THE YORKSHIRE WATER NETWORK SIMULATION MODEL

```
┌──────────────────── MAIN MENU ────────────────────┐
│     1   Run network simulation model               │
│     2   Run reservoir simulation model             │
│     3   Run pumping cost model                      │
│     4   Run control line program                    │
│     5   Edit reservoir data                         │
│     6   Edit network data                           │
│     7   Edit cost data                              │
│     8   Estimate costs                              │
│     9   View grid                                   │
│    10   Print results                               │
│    11   Symphony                                    │
│                                                     │
└────────── Press ENTER to continue or ESC to return to DOS ──────────┘
```

8.1 (line 4). Running the control line program calculates the water levels in reservoirs above and below which water can be abstracted at set rates and yet not run dry. If we choose a particular reservoir (say Broomhead), the model first estimates the amount of rainfall which can be expected in a given time period. This is based on historical time series records. The model is thus able to estimate the amount of inflow to the reservoir. The next operation is to calculate the minimum inflow that could occur in a 24-month

Table 8.2 Results of model run with increased demand (*Source*: GMAP 1993)

Demand node	Deficit after balancing (TCMD.)*		For a complete set of results see the following files:	
	1960	1976		
1. Bradford	0	0	GDRY.FLO :	Average flow through network in 1960
2. Lower Wharfedale	0	0		
3. Embsay	0	0	GWET.FLO :	Average flow through network in 1976
4. Harrogate & Ripon	0	0		
5. Leeds	0	0		
6. Calder	0	0	GWEEK.FLO :	Flow through network in week 100
7. Denby Dale & P'stone	0	0		
	0	0		
The model shows us if there are any average deficits at the demand points in the two chosen years. It tells us in which files we can find more information. PRESS ENTER	0	0	ELVHARD. DAT :	Minimum hardness at Elvington
	0	0		
	0	0		
	0	0		
	0	0	• Negative deficit = surplus	

sequence, as in a particular severe drought period. From these calculations the model sets target abstraction rates, and compensations that may be required from other reservoirs, in order to preserve supplies. First it estimates the 'normal control line' – the minimum level of water required in a reservoir to maintain normal abstraction rates. (The difficulty for water companies is that these are generally lower in winter, when rainfall is higher, because of the seasonal demands for water from consumers.) If levels fall below this line, new drought rates of abstraction come into force. Second, it can estimate the 'drought control line' – the minimum level of water in a reservoir needed to maintain drought abstraction rates. If levels are estimated to fall below this line, emergency procedures must be adopted. This procedure can be undertaken for all reservoirs using the 'reservoir simulation model' option shown in line 2 of Table 8.1.

A second major use of the model is to simulate the impacts of changing levels of demand. Suppose, for example, an area had a considerable input of new houses and it was wished to estimate the likely impacts on demand

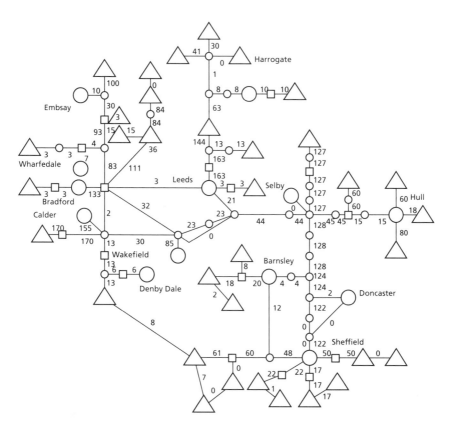

Figure 8.4 Allocation of supply through the grid
(*Source:* GMAP 1993)

on certain parts of the network. The 'network simulation model' can now be run (from line 1 of the opening menu shown in Table 8.1). With the new demand, the model will allocate available water from all supply points to the new set of demand points via the grid system. It could be assumed conditions apply to a wet or dry year, and one can choose any part of the year to understand annual variations in demand. If a particular week is chosen in a year, the model will pinpoint any deficiencies in supply that the new demand may create. If the new demand cannot be satisfied from existing supply, the flows through the grid would have to be adjusted accordingly. Table 8.2 shows the results of a model run for conditions which replicate a dry year (1960) and a wet year (1976). For this particular increase in demand, there are no problems supplying that overall demand by week 100 (of the two-year simulation cycle); hence the rather dull but reassuring set of zeros in Table 8.2. We can plot the allocation of supply through the grid network in graphic mode (Figure 8.4).

Having calculated flows through the grid given certain new conditions, the costs involved can be estimated. Costs are generally made up of pumping costs, treatment costs, and electricity tariffs. If changes are known to have taken place in any of these rates, these can be included at this juncture. Costs can then be estimated for each link in the grid and displayed in graphical form (Figure 8.5).

8.2.2 Water quality

The concern caused by environmental damage to rivers and coastlines through pollution and contamination has been a major incentive to invest in computer support to monitor outbreaks and diffusions. As with most GIS developments, water-quality monitoring began with simple mapping packages to visualize the relationships between incidences of contamination and possible source locations. Keith Annand (interviewed in Thomasson 1989) of the NRA Thames Region describes the 'Tamesis' information and mapping system (Thames Asset Management and Environmental Services Information System) produced in the late 1980s. Tamesis provides:

> complete map coverage, the entire river network, all the statutory flood plans, a complete database of the sites of specific scientific interest, locational and postcode information, all base population statistics and digitized subcatchments of the river.
>
> (Thomasson 1989: 38)

The importance of researching the links between potential contamination and 'populations at risk' has been recognized for some time in the literature on water-quality issues. Table 8.3 shows the possible sources of such contamination at various stages in the water cycle. Some of these may occur throughout the cycle, others are more localized. Examples of the former

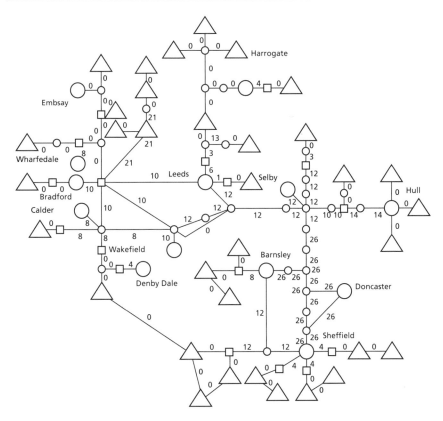

Figure 8.5 Estimation of costs along the grid
(*Source:* GMAP 1993)

would include the general concerns in many urban areas of the effects of adding fluoride to water supplies in the late 1970s and early 1980s. This initiated a number of studies investigating the links between fluoride and cancers. More localized occurrences of contamination are more likely to be linked to a particular source, although often unequivocal proof is difficult to ascertain.

It is clear that the mapping capabilities of GIS would be useful in highlighting spatial incidences of possible contamination. However, the information needs are much broader in total. Since water companies have an increasing need to hold, display, analyse, and report data resulting from responsibilities to maintain and improve water quality, they are continually seeking to develop comprehensive information systems that include water-quality information. The key requirements include those to:

- display and analyse key data by specified areas;
- monitor performance against regulatory standards;

Table 8.3 Possible contamination sources at various stages in the water cycle (*Source*: GMAP 1993)

Pollution source:
Sewage treatment works (STW)
Storm sewage overflows (SSO)
STW and SSO
Urban run-off
Industrial discharges to river
Landfill leachates
Minewaters
Agriculture
Other

- update and amend data and regulatory changes;
- map the current environmental burdens;
- model and map future environmental burdens of regulatory change.

One system already in operation is a potable water quality information system (PWQIS) designed by a consortium including Yorkshire Water, the University of Leeds, and the NRA (see McDonald et al 1993). The structure of the information system is shown in Figure 8.6. There are three essential elements. The first is the database, which contains the results of the water-quality determinations at individual sites. There are, typically, some 50 000 determinations per year in a large water plc for approximately 60 parameters. The key parameters, some 15 of which are subject to failure, are held within this database (McDonald et al 1993). New water-quality determinants are added every three months through a data appending system. A second core element of the database holds the current UK standards, known as the prescribed concentration value (PCV). The final element is a set of digital boundary files, which identify the spatial characteristics of each of the possible reporting areas together with information on the structure of the spatial hierarchy.

The second major element of the system is the analytical engine. Its first operation is to compare PCVs with appropriate water-quality determinations to evaluate the characteristics of failures in water quality (number, frequency, distribution, location, etc.). The second step of the model tests for relaxations and undertakings, which will reduce the failures by recognizing 'expected failures' for which legal provision has been made. The third stage identifies and excludes minor infringements to leave only those sites containing unprotected failures.

The final major element of PWQIS relates to output structures. These may be provided for every parameter by area and by three-month period. The system provides summary tables (Table 8.4), statistical tables (Table 8.5), and mapping capabilities (Figure 8.7). Comparison of period-by-period results

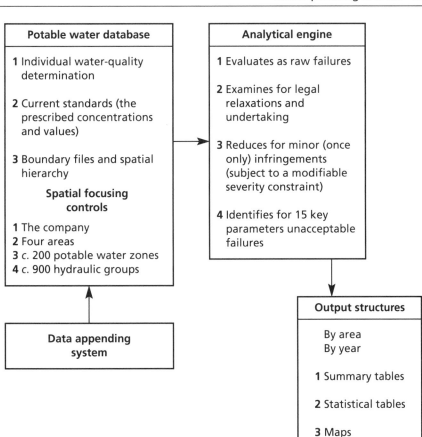

Figure 8.6 Structure of PWQIS
(*Source*: GMAP 1993)

identifies regional water-quality trends, which may be in part associated with raw water deterioration stemming from environmental change. Treatment works established or expanded to deal with raw water deterioration will have an environmental by-product, the location and scale of which can be estimated in advance through PWQIS.

An interesting application lies in the evaluation of the impact of environmental legislation. PWQIS can be fed scenarios of possible new prescribed concentrations and values to provide most likely 'what-if' solutions. Existing data can then be re-evaluated to determine the extent and severity of failures that would be induced by new legislation (and the consequent additional treatment facilities and by-product that would stem from this). For example, if the current concentration limit for aluminium, 200 mg/l, was changed to 50 mg/l, it would be possible to determine whether this would impact equally upon all areas within a water company or, potentially, upon all

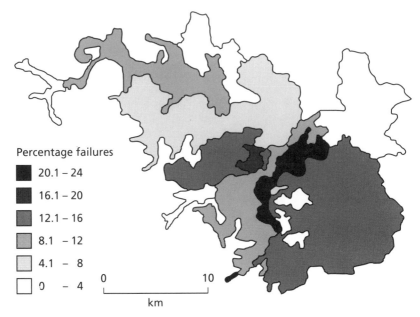

Percentage failures

- ▇ 20.1 – 24
- ▇ 16.1 – 20
- ▇ 12.1 – 16
- ▨ 8.1 – 12
- ▨ 4.1 – 8
- ☐ 0 – 4

0   10
km

Figure 8.7 Mapping capabilities in PWQIS
(*Source*: GMAP 1993)

water companies in a European member nation, or indeed the equity in the spread of the impact of the regulation throughout all member countries.

A second example of integrating GIS and spatial models is provided by the CATNAP system, of particular relevance to the work of the NRA. As discussed earlier, the NRA uses a system of consents through which the

Table 8.4 Summary tables from PWQIS (*Source*: GMAP 1993)

	Failures	Insignificant	Enforceable	Undertaking	Relaxation
AL	0	F	F	F	F
FE	3	F	T	T	F
MN	5	F	T	F	F
Colour	0	F	F	F	F
TURB	0	F	F	F	F
PAH	0	F	F	F	F
NO3	0	F	F	F	F
NO2	0	F	F	F	F
THM	1	T	F	F	F
TotColi	1	T	F	F	F
F. Coli	0	F	F	F	F
Pesticide	0	F	F	F	F
PB	0	F	F	T	F
PH	0	F	F	F	F
Other	0	F	F	F	F

F = False (no activity recorded)
T = True (activity recorded)

Table 8.5 Summary statistics from PWQIS (*Source*: GMAP 1993)

	Samples taken	No. fails	% fails	Min.	Mean	Max.	Prescribed concentration value	Exceedence	Enforceable
AL	97	2	2.06	34.00	85.56	340.00	20.00	140.00	F
FE	96	8	8.33	12.00	92.05	620.00	60.00	420.00	F
MN	97	11	11.34	5.00	24.73	220.00	4.00	170.00	T
Colour	103	0	0.00	2.00	3.89	10.10	5.5-9.5	0.00	F
TURB	103	1	0.97	0.10	0.51	8.70	200.00	4.70	F
PAH	7	0	0.00	0.02	0.02	0.02	200.00	0.00	F
NO3	35	0	0.00	2.20	2.35	4.50	50.00	0.00	F
NO2	85	0	0.00	0.03	0.03	0.10	50.00	0.00	F
THM	7	0	0.00	3.70	17.59	49.10	50.00	0.00	F
TotColi	103	2	1.94	0.00	0.00	6.00	0.10	6.00	F
F.Coli	103	0	0.00	0.00	0.00	0.00	0.20	0.00	F
Pesticide	0	0	0.00	0.00	0.00	0.00	100.00	0.00	F
PB	90	1	1.11	5.00	8.22	127.00	10.00	77.00	F
PH	103	0	0.00	6.90	8.71	9.50	0.00	0.00	F
OTHER	0	0	0.00	0.00	0.00	0.00	0.00	0.00	F

discharge of pollutants into, and the abstraction of water from, the river system are controlled. Although national discharge consent standards do not exist, the NRA has set itself clear attainment targets, and its performance in attaining the targets are monitored, particularly by using catchment water-quality models. CATNAP contains processes of mass balance, reaeration, self-purification, and thermal exchange. It allows a river network to be defined for a study area as a series of interlinked reaches, each with its own hydraulic and process characteristics. The speed of execution allows a realistic 'what-if' capability; a powerful visual presentation module integrates digitized river catchment networks with model results and provides graphical and summary output. (Legislation demands a statistical basis to the evaluation for consents, and this must be provided in addition to the basic GIS display.)

The model proceeds by defining a reach in terms of its length and channel geometry. It can then be attached to other single reaches in the system, or a confluence or bifurcation can be defined as appropriate. Flows are defined for both the head of the main river and any tributaries, and for lateral inflow to each reach. Process characteristics are then defined for each reach, covering self-purification, reaeration, and thermal exchange, along with information defining the flow – travel-time relationship for that reach. For example, estimates of water velocity can be made in a variety of ways. As this controls the time of degradation of pollutants in a reach, it is important that users can explore the different effects of various assumptions on water velocity. In addition, the quality of lateral inflow water for each reach can also be defined. To simplify the task of providing a large quantity of catchment data, standard sub-catchments can be defined for both quantity and quality. These can then be picked up and proportioned for use where appropriate.

To calculate the amount of self-purification or reaeration, the travel time to the next feature is first estimated. Then first-order expressions, incorporating temperature correction, are used to determine the amount of self-purification which has occurred. The model incorporates a form of the Streeter-Phelps equation to describe the oxygen balance, and includes the equivalent oxygen demand of ammoniac nitrogen. In addition to the above oxygen exchange processes, the model also incorporates the effects of reaeration over weirs. The air–water temperature exchange is modelled on a simple first-order basis.

The model allows representation of abstractions and discharges to be made in a number of ways. For abstractions, the amount of water taken can be specifically defined. In addition, the model has the ability to carry out abstractions subject to prescribed flow restrictions in the river. Abstracted water can be returned at one or several locations to the river, and allowances can be made for its partial or total loss. The quality of the returned water, for each determinant, can be specified in a number of ways. It can either reflect directly the water taken out, be subject to a change based on the abstracted water, or be completely re-specified anew. Similarly, the system provides flexibility in investigating the impact of discharges. For the flow component of any discharge, the minimum and maximum flows, based on the original distribution, can be specified. This allows for the easy representation of several effluent streams from a plant, which come into or out of operation at varying flows. To extend this further, flows and qualities from a discharge can be correlated. This allows, first, for several discharges from one plant to be synchronized. Secondly, processes which are related or inversely related to flow can also be correctly represented. By the use of scale factors, the future increase or decrease in magnitude of a discharge can be represented.

To allow scenarios to be investigated for discharges, the system first allows for a discharge performance to be specified. This permits a quick examination of the impact of discharging at a particular consent. For a scenario to be examined in further detail, a range of performance values can be specified, and the results of each are produced. The model allows for either original results or revised performance to be carried forward.

To enable the model to be calibrated, data from flow- and quality-monitoring stations can be taken into the model and compared with predictions, using the Kolmogorov-Smirnov statistical test. This test, which can specify the determinants to be tested, is carried out across the whole of the probability distribution. Due account is taken of the number of data items representing the observed distribution. The model runs can be automatically refined to carry out a flow or quality calibration only, or both together. In addition, a selection can be made to include either all data or natural data only. This is particularly useful in establishing the natural hydrological flow balance for a catchment.

This model and its predecessors have been used in Yorkshire for nearly ten years in determining the impact of continuous discharges on receiving waters. With this new modelling system, it is now much easier to build models for a particular catchment, and verify that data have been entered correctly, by use of the input data facilities and visual representation of information. In addition, this visual representation makes the task of explaining the impact of a proposal to managers or other interested people much more straightforward. It is now envisaged that this model will be widely used in Yorkshire for carrying out such assessments, and the knowledge gained will be used to refine further the interaction and visualization capabilities of this tool.

Again, to be more specific about the context and uses of CATNAP, the river water quality in two Yorkshire catchments (the Aire and Calder Valleys) can be discussed. These catchments have serious environmental problems. Below Leeds, for example, the dry-weather flow of the Aire is only 30 per cent natural, with 70 per cent of flow derived from effluent. The broad water-quality objectives for the river as defined by the NRA are shown in Table 8.6. That these objectives cannot be attained at present is attributable to many causes, the principal of which are shown in Table 8.7. Many different strategies and mixes of strategies may be employed to remedy the situation. For example, it is possible to reduce effluent volume, improve effluent quality, alter effluent input sites, create oxidation lakes, introduce oxygen curtains, develop inter-basin transfers for dilution, and so on. CATNAP replicates and demonstrates the current fluctuations in water quality in the basin and simulates the basin response to various remedial strategies as expressed in new consent limits and river conditions. For more details see McDonald et al (1993) and Clarke et al (1993).

8.2.3 Waste-water disposal

One of the most effective ways in which modelling technology can become valuable is by allowing issues of 'change' to be addressed. In this section, we will describe a typical application to a pressing policy issue in the water industry. The issue in question is an EC directive to control waste water ('sludge') disposal into Europe's seas. At present, the North Sea is used as a primary waste-water outlet by Yorkshire Water. The company will therefore have to respond to the new legislation by identifying alternative disposal strategies.

One alternative strategy is to spread more sludge on the land. This approach can then be used to tie in with another EC-driven policy of 'set-aside', in which

Table 8.6 Water-quality objectives on the River Aire (*Source*: GMAP 1993)

River and class		1989		Objective	
		(km)	*(%)*	*(km)*	*(%)*
Aire	1a	110.5	23.3	137.3	28.9
	1b	146.4	30.8	165.0	34.7
	2	72.4	15.2	169.3	35.6
	3	129.2	27.2	3.4	0.7
	4	16.5	3.5		
Calder	1a	102.4	24.1	118.3	27.8
	1b	99.0	23.2	129.0	30.3
	2	106.3	25.0	161.0	37.8
	3	98.3	23.1	17.3	4.1
	4	19.6	4.6		
Canals	1a				
	1b	66.0	33.6	121.7	61.9
	2	77.3	39.3	73.7	37.4
	3	53.4	27.1	1.3	0.7
Total	*1a*	*212.9*	*19.4*	*255.6*	*23.3*
	1b	*311.4*	*28.4*	*415.7*	*37.9*
	2	*256.0*	*23.3*	*404.0*	*36.8*
	3	*280.9*	*25.6*	*22.0*	*2.0*
	4	*36.1*	*3.3*	*22.0*	*2.0*
Grand total		*1097.3*	*100*	*1097.3*	*100*

parcels of agricultural land are freed from productive use in order to cut down on surpluses. The problem is, however, further complicated by the fact that there are more regulations which govern the amounts of sludge which can be spread on the land. These regulations do not relate to the actual mass of the sludge, but rather to the concentrations of toxic chemicals such as cadmium and zinc which the sludge contains. For example, sludge may only be spread on the land up to a toxicity threshold of 3 g per hectare.

Table 8.7 Reasons why objectives cannot be attained (*Source*: GMAP 1993)

Pollution source	Aire	Calder
Sewage treatment works (STW)	48.3	45.8
Storm sewage overflows (SSO)	21.0	9.7
STW and SSO	11.8	10.0
Urban run-off	5.1	4.2
Industrial discharges to river	1.8	0.6
Landfill leachates	2.3	9.4
Minewaters	1.6	12.1
Agriculture	2.7	1.8
Other	5.4	6.4

Table 8.8 Outputs from treatment works in Yorkshire available for sludge disposal (*Source*: GMAP 1994)

Waste water treatment works data				
WWTWs	*Treated*	*Pop(A)*	*Toxicity indicator (Zn)*	*Toxicity indicator (Cu)*
Mitchell Laithes	5482.84	124600	0.00	0.00
Old Whittington	2811.49	118810	1421.78	1722.57
Naburn	2419.68	79776	455.00	440.00
North Bierley	2403.35	41609	670.75	446.50
Aldwarke	2323.26	92210	810.64	369.50
Caldervale	2302.07	91860	909.18	441.27
Woodhouse Mill	1942.11	92320	1118.85	544.95
Sandall	1924.96	85090	730.29	509.93
Marley	1893.65	76660	0.00	0.00
Spenborough	1758.47	48634	0.00	0.00
Lundwood	1663.34	81530	877.47	346.60
Beverley	1006.87	28030	316.00	215.00
Sutton	865.51	42480	902.67	489.33
Dowley Gap	849.48	33580	3436.67	172.67
Harrogate South	838.57	35050	5127.50	841.00
Thorne	829.68	38750	574.00	230.40
Rodley	807.34	42290	0.00	0.00
Mill Lane	777.31	36007	1696.67	300.00
Ossett Spa	753.68	16470	502.75	277.25
Harrogate North	692.45	38530	363.75	116.53
Neiley	630.06	16120	769.00	185.00
Owlwood	610.97	32413	374.00	325.00
Bransholme	610.77	34150	350.00	370.00
Wombwell	604.80	33800	0.00	0.00

There are two further complications to consider in trying to decide how to spread sludge waste. The first is that different treatment sites will produce waste output with different toxicity profiles. Clearly, if one works produces sludge with a zinc concentration of 6 g of zinc per tonne, then only half a tonne of this sludge can be applied to a hectare of land. On the other hand, if another works produces sludge with a zinc concentration of 2 g per tonne, then it is possible to apply one-and-a-half tonnes to a hectare. The second complication is that certain land types have either fully or partially restricted thresholds of toxicity: these land types include sites of special scientific interest (SSSIs), aquifers, boreholes, and, of course, urban areas.

To address this problem formally, one can begin by identifying the location of all treatment works in the Yorkshire region, each of which has an estimated level of production of treated output with an associated toxicity profile. A sample of these outputs is shown in Table 8.8. Next there is a need to identify potential land parcels for the receipt of treated wastes. In effect,

Table 8.9 Allocation of wastes from treatment plant (*Source*: GMAP 1994)

	Land data	
Land	*(2) ARBRKT* *Sludge Src*	*(2) ARBRKT* *Sludge Alloc*
482512	East Barnby	1.43
431503	East Cowton	13.24
493477	East Heslerton	6.26
391450	East Marton	4.02
397426	Eastwood	83.25
397427	Eastwood	83.25
398426	Eastwood	83.25
396426	Eastwood	83.25
396427	Eastwood	83.25
397425	Eastwood	72.95
458425	Eggborough	75.24
481505	Egton Bridge	8.01
442475	Eldmire	105.41
491428	Ellerker	117.54
471440	Ellerton	1.79
440401	Elsecar	140.00
440400	Elsecar	138.04
469448	Elvington	8.80
399453	Embsay	36.04
462442	Escrick	11.09
427396	Ewden Village	5.51
468489	Fadmoor/Gillamoor	3.58
461464	Farlington	1.25
420481	Fearby	1.84
447485	Felixkirk	1.25

these parcels can be identified by conventional overlay techniques which progressively exclude urban land, sites of special scientific interest, and so on. An example is shown in Figure 8.8 (Plate 8.2). The problem is then to devise an allocation of wastes from treatment works to available land parcels in a cost-efficient manner (in practice, by minimizing the cost of transportation of wastes). Table 8.9 shows a selection of the resultant allocations, in which individual kilometre squares are identified in the left-hand column of the table, a treatment works is identified in the central column, and the size of the allocation (in tonnes) is shown on the right.

The uses to which lands under sludge may be put are clearly somewhat restricted: the most likely outcome is for 'short-rotation coppice' (that is, fast-growing trees) to be grown. These trees may be incinerated to produce electrical power. A second component to the modelling problem is to identify suitable (optimal) locations for new incineration plants ('gassifiers'), again with the objective of minimizing the costs of transferring biomass from the place of production to its place of incineration. Figure 8.9 (Plate 8.3) and Table 8.10 show examples of biomass inputs to a proposed new gassifier at Eggborough in south Yorkshire.

8.2.4 Water-demand forecasting

So far this chapter has concentrated on the supply side of water provision, putting the case for GIS and modelling systems aimed to explore 'what-if' questions concerning future water qualities, reservoir quantities, and the

Table 8.10 Biomass inputs into new gassifier at Eggborough (*Source*: GMAP 1994)

Land data	
Land	(2) ARBRKT DistGen
458425	1.00
460427	3.66
462423	4.22
454420	5.59
455419	5.78
456418	6.29
451424	6.90
451425	6.97
452428	7.13
452426	7.18
451427	7.52
465429	8.68
458433	9.00
457433	9.04
464432	10.06
453433	10.25
467430	10.90
469427	11.50
462435	11.74
470423	12.14
470428	12.74
467415	12.80
446419	12.91
471425	13.14
447416	13.52
467414	13.52
468415	13.53
447414	14.79
472429	14.96

ability of networks to cope with 'worst-case' scenarios. Water is supplied for domestic (household) and industrial consumption, yet very little accurate information exists on the variations in water demand across cities and regions. Whilst industrial premises are metered in the UK, only 3–5 per cent of household water consumption is recorded in this way. Water companies are increasingly interested in the forecasting kitbag that contemporary quantitative geographers have at their disposal.

Water-demand forecasting has attracted the interest of geographers and economists for some time. Grima (1971) is an early example of a detailed attempt to calculate household demand across a city and to test alternative mechanisms for charging for water usage. In one sense, the arguments that follow in this section are an attempt to update this type of micro-scale research. Unfortunately, many of the attempts since Grima (1971) have been based on regression techniques, where the interest has been at the city- or region-wide scale. A typical model to predict water demand would take the following form:

$$Y = f(X_1, X_2, X_3) \qquad\qquad (8.1)$$

where the variables might be for example:

X_1 = average summer temperature;
X_2 = characteristics of the household (size, age);
X_3 = perhaps an exogenous variable such as price of water.

Obviously this is a gross simplification, and more detailed models can be found in Batchelor (1975), Gardiner and Herrington (1986), and Danielson (1979). The skill in this modelling procedure is to estimate the coefficients associated with each of these variables, in order to calibrate the models to real-world data.

Many of these factors are clearly more favoured in regions that experience wide climatic variation from one end to the other. Any attempt to disaggregate these models further usually breaks down because of problems regarding data availability.

This section outlines the usefulness of microsimulation methods for the estimation of household water demand and consumption, the subject of a major research programme at Leeds sponsored by Yorkshire Water. As we saw in Chapter 7, microsimulation is a technique for estimating micro-level data using chain conditional probabilities. The use of conditional probabilities allows the incorporation of the widest range of available known data to reconstruct detailed micro-level populations.

Microsimulation models offer considerable potential in the modelling of household dynamics. Given the rapid dating of Census information in a city or region, coupled with genuine worries over the future of the Census itself, this is crucially important. On the demand side, the main principle is to update individual and household attributes by means of list processing. This involves deriving conditional probabilities for events such as deaths, births, migrations, etc., and to invoke Monte Carlo sampling methods to determine whether eligible individuals undergo appropriate transitions (Clarke 1986, Rees et al 1987, Duley et al 1988).

There are a number of reasons why this is an important time for undertaking such research. First, as stated above, with the relative absence of domestic metering in the UK, a detailed database on household water consumption is lacking for cities and regions. Although there are estimates of the supply of water at various junction points within a city, there is no knowledge of the real spatial variations. Following privatization, the water industry has been forced to examine the costs and benefits of future investment decisions in a far more accountable fashion. It desperately needs better information on current and future consumption patterns.

Secondly, there is increasing interest in what the implications of small-area variations of consumption might mean for future pricing policies. Where meters have been installed it is clear that overall household consumptions do decline (for instance, in the Isle of Wight), but there is not much evidence of who gains and who losses, especially when it is revealed that many water companies participating in metering trials have deliberately kept prices lower in order not to upset the 'guineapigs' in the trial areas. In addition, if household consumption patterns are estimated for a city, it may be possible to judge the necessity for metering and the effectiveness of that metering. This could be done by targeting metering at those areas where the water authorities would gain most benefit. The simulation exercise could look at the effectiveness of alternative, household-based pricing policies. The pricing mechanism for water has traditionally been based on the rateable value of a property, ignoring quantity used or the actual income of present occupiers. This situation is required by law to end in the UK sometime in the late 1990s. The obvious question is which pricing mechanism will replace the rateable value system, given both the costs of installing meters in every property in the UK and the controversy caused by large, poorer families having to pay new, very large water bills. Any move to change pricing policies would of course be politically sensitive (we have seen this clearly in changes to the UK domestic rate system). However, the simulation exercise would allow a variety of future scenarios concerning pricing policies based on actual consumption or income to be tested and evaluated in terms of likely gainers/losers. Thus, it would be interesting to work towards pricing scenarios which might benefit all water users by adopting flexible policies, which might then vary across a city.

Although the water industry rarely publishes its methods for estimating domestic consumption, it is clear that geodemographic systems are in

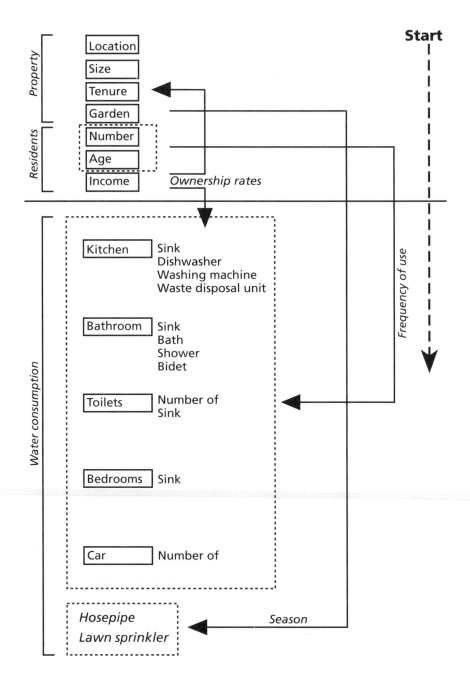

Figure 8.10 Structure of water-demand simulation model
(*Source*: Clarke et al 1995)

operation (cf. Chapter 3). In the UK the Severn Trent Authority has a large metered sample of households which it aggregates into ACORN groups and sells to other regional authorities. By purchasing ACORN data for their regions, different authorities can apply the typical water consumption patterns of these various Severn Trent ACORN groups in order to estimate small-area demands in their localities. As in all geodemographics, this will work well in some areas, poorly in others. In the case of water consumption, the type of house is not as important as the number of residents within it and the amount of water appliances they might actually have. Russac et al (1991: 350) looked at ACORN profiles for a sample of metered houses in Brookmans Park, north London. They concluded: 'There are significantly different consumption figures for the same ACORN classifications and also that similar consumptions occur with different ACORN groups.'

The alternative to geodemographics is to build up detailed household profiles containing information related to incomes, family size, and likely ownership of water-using appliances. There is a detailed literature on the characteristics which influence levels of consumption (see Gardiner and Herrington 1986, Danielson 1979, Batchelor 1975). Figure 8.10 shows the types

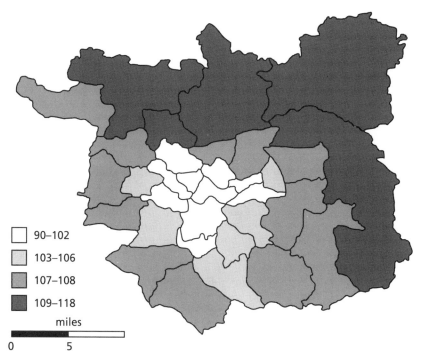

90–102

103–106

107–108

109–118

miles

0   5

Figure 8.11 Household water demand in Leeds: by property type
(*Source*: Clarke et al 1995)

of individual and household variable which are most likely to influence both the ownership rates of water-using appliances (normally income-related) and the frequency of use of such appliances (household size variables). The diagram also sets out the likely steps in the simulation exercise. The important point is that by considering only one of these, a very different pattern of consumption becomes evident. Figures 8.11–8.13 show the average annual household water demand for wards in Leeds by property type, tenure type, and number of occupants respectively. These are all calculated directly from raw data sets at the household level. The patterns are clearly not the same. That is, the average, and hence total, household consumption patterns will vary according to the categorization type. The task in the microsimulation exercise is to estimate the joint and conditional probabilities of household water usage based on a combination of all these factors.

The simulation model can be expressed formally as:

$$X_k^r = X_k^r \left[P_k \left(X_k^r \text{ I...} \right) R_i^k \Gamma \right] \tag{8.2}$$

where:

P_k are conditional probability distributions;

R_i^k is a random number;

Γ represents a set of constraints.

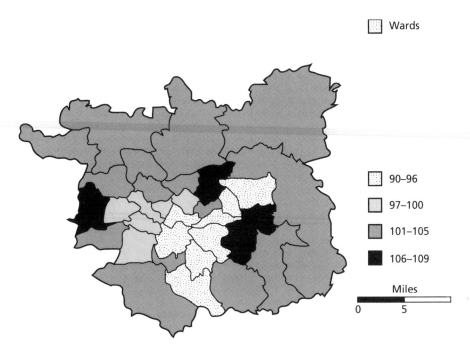

☐ Wards

☐ 90–96

☐ 97–100

▨ 101–105

■ 106–109

Miles

0   5

Figure 8.12 Household water demand in Leeds: by tenure type
(*Source*: Clarke et al 1995)

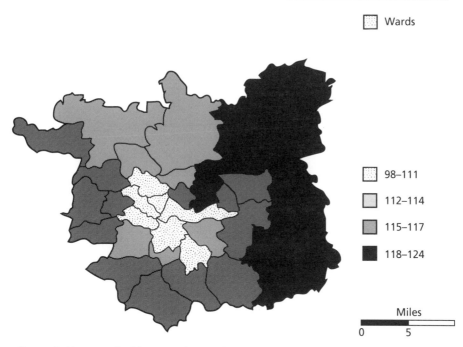

Wards

98–111

112–114

115–117

118–124

Miles

0   5

Figure 8.13 Household water demand in Leeds: by number of occupants
(*Source*: Clarke et al 1995)

The probability distribution is divided into intervals (on a scale from 0 to 1) each one corresponding to a value of X_k^r. Random number tables can then be used to 'select' an interval, and hence X_k^r is assigned the corresponding value. This procedure generates a population with characteristics such that they satisfy all known probability distributions (see Clarke and Holm 1987, Birkin and Clarke 1988).

The first step in the procedure is to ascertain the location of the property to be used in the sample, which may be all households in a small area or a selection of households across the entire city. Given the location of the household, the likely household size and tenure and the characteristics of the tenants within can be estimated (see Birkin and Clarke 1988 for such a procedure within Leeds). This in effect then gives us the top half of Figure 8.10. From here the likely ownership rates of water-using appliances within different households (washing machine, dishwasher, etc.) can be estimated by combining the simulated sample with information contained within the General Household Survey. The household size variables then allow the calculation of the frequency of use. Since the water industry holds detailed information concerning the amount of water used by different appliances, frequency of use can be translated into actual consumption patterns. Combining income and household size variables in this way thus gives detailed small-area estimates of water demand for the first time in the

UK. In order to calibrate the models there are a number of additional factors which help:

- the known household characteristics of EDs: these allow us to reproduce the household size and income-related characteristics of our sample confidently (as in Birkin and Clarke 1988, 1989);
- the information from those properties which have been metered: this allows us to look in more detail at the relationships between household characteristics and water consumption;
- the known supply totals at various supply junctions across a city (though leakage has to be borne in mind here).

There are a number of advantages to the procedure outlined above. The first relates to coverage and interdependency. In this routine, one does not have to make average assumptions about households in a particular neighbourhood. The interdependencies between income-related variables and household size and location can be modelled directly. Secondly, flexibility over the aggregation process can be retained, so that the household information can be combined into any convenient spatial zones (Census wards or postal districts, for example). Thirdly, the likely consequences of changes to any of the variables shown in Figure 8.10 can be explored. The basic population set might be augmented by the construction of a new housing estate; the older inner-city residences might be 'yuppified' into high-class unit accommodation; average household size might steadily decline over the next decade; new technologies might impact on the efficiency of water use by many standard appliances. All of these scenarios could be simulated in the modelling procedure. More fundamental, perhaps, is the crucial advantage of microsimulation over regression techniques in its ability to update the synthetic data set directly through list processing. This involves deriving conditional probabilities for events such as birth, death, migration, etc., and testing whether individuals undergo such transitions through Monte Carlo sampling methods. The advantage of microsimulation in household dynamics relates to the handling of the interdependencies between an individual's attributes and between individual members of households (Birkin and Clarke 1988, Clarke 1986).

Fuller details of the issues raised above appear in Clarke et al (1995).

8.3 CONCLUDING COMMENTS

The water-supply system clearly involves a series of complex interactions between hydrology, engineering, economics and geography. It is also an emotive area as far as the general public are to concerned as the problems of water supply caused by the UK drought in 1995 have re-emphasized. They expect a cheap but reliable service, and place considerable pressures on the water regulators to ensure such expectations are met. For these reasons it is not surprising that planners in the industry have been keen to invest in more sophisticated spatial analysis tools to address problems related to water quality, water demand, leakage, and general network planning. We hope that this chapter has illustrated the considerable *applied* usefulness of GIS and spatial modelling for water planning. The systems described above all provide customized, practical, computer-aided modelling frameworks to aid in the formulation and application of strategic environmental planning and decision making, with the help of advanced GIS techniques. That all of these systems have (in part) been commissioned and sponsored by a leading water company is testimony to their applied importance and usefulness.

CHAPTER 9

SDSS for the motor industry

9.1 INTRODUCTION

The automobile industry spans a vast range of vehicle and component manufacturers, retailers, and distributors, as well as finance and insurance companies. In the European Community it accounts for approximately 9 per cent of GDP and 10 per cent of total employment. The car is the ultimate consumer product of the twentieth century – in turn liberating individuals' personal mobility and generating a whole set of political, economic, and environmental debates. Cars inspire affection and aspiration, loathing and selfishness. A fascinating insight into the 'machine that changed the world' is provided by Womack et al (1990).

The car has changed the geographical dimensions of modern life in a way that could not have been envisaged 70 years ago. Now 80 per cent of all US households own at least one car, with the average being almost two cars per car-owning household. In the UK, where car ownership is less well developed, currently 66 per cent of households are car-owning. The Department of Transport in 1993 projected a growth from the existing 21 million cars on the road to 51 million in the year 2010 given existing trends. As Table 9.1 shows, internationally there is a direct correlation between prosperity and car ownership levels. Whatever the impact of environmental legislation or national transport policy, the only realistic prospect is for a significant increase in the car population of the UK and most developed countries.

Despite these increases in car ownership rates the motor industry, like most other consumer product markets, is subject to the impacts of fluctuating fortunes in the economy. During the late 1980s the total number of new car registrations in the UK hit record levels as the economic boom fuelled demand for consumer products. Table 9.2 shows the annual sales of new cars in the UK in the late 1980s and early 1990s. There is probably no other large consumer market that shows such market size volatility over relatively short time spans.

The overall market slump identified in Table 9.2 is reflected also in the performances of individual manufacturers. For example, between 1990 and 1991, Ford's new car sales fell in the UK from 507 000 new cars to 385 944, culminating in a fall in market share from 25.3 per cent to 24.2 per cent and an operating loss of £761 million. Levels of sales and market share are key

Table 9.1 Prosperity and car ownership levels (*Source*: adapted from the *Financial Times* 1994)

	GDP per capita ($)	Persons per unit owned
USA	24 000	2.0
Switzerland	22 000	4.0
Germany	20 000	3.8
Japan	19.000	4.0
Canada	19 000	3.0
France	18 000	4.0
Italy	16 500	3.8
Australia	16 000	3.0
Singapore	15 500	9.5
UK	15 400	4.5
Saudi Arabia	11 000	7.5
Taiwan	10 000	10.0
Korea	9 000	20.0
Malaysia	7 000	17.0
Mexico	6 500	16.0
Thailand	5 500	50.0
Russia	5 500	15.0
Brazil	5 000	18.0
Turkey	4 500	45.0
Indonesia	3 000	90.0
Philippines	2 500	85.0
China	2 500	400.0
India	2 000	300.0

drivers of profitability and therefore a major concern of senior executives in all manufacturing companies.

In addition to an overall decline in total registrations, the industry itself is facing major changes. The overall market is becoming more competitive as a result of a number of processes. First, internal restructuring has produced massive increases in productivity for a number of companies. The Monopolies and Mergers Commission (1992) estimate that Vauxhall and Peugeot were able to increase productivity by 75 per cent between 1986 and 1990. Second, the market share of Japanese manufacturers has increased at the expense of European manufacturers. In the UK the Japanese are quota restricted to about 11 per cent of the overall market. However, with the arrival of manufacturing plants in the UK (Nissan at Sunderland, Toyota at Derby, and Honda in Swindon), the Japanese have been able to bypass the EC import quota restrictions and retail on a much more aggressive basis. Again, the Monopoly and Mergers Commission (1992) estimate that by 1999 Japanese sales will have reached 500 000 or approximately 25 per cent of the market (compared with 12 per cent at the time of writing). If the Commission's recommendations that 'voluntary export restraints' between the EC and Japan are removed more quickly are followed, greater

Table 9.2 Total UK new car registrations
1986–92 (*Source*: GMAP 1993)

Year	Millions
1986	1.88
1987	2.01
1988	2.22
1989	2.30
1990	2.01
1991	1.59
1992	1.62

competition will create more Japanese sales. Third, it is likely that the number of major players in the market will reduce as power is concentrated into the hands of fewer manufacturers. This has been a feature of most other retail markets during the 1970s and 1980s (Kay 1987), and the first signs are emerging in the car industry; for example, with the coming together of Saab and General Motors and with the proposed merger in 1993 of Renault and Volvo. In fact this merger never occurred due to differences in the views of the shareholders of both companies. Manufacturers are emerging as either volume-led, global companies addressing mass markets (Ford, Toyota, GM, Nissan) or more specialist, niche manufacturers aiming for profitable penetrations of specialist markets (BMW, Mercedes, Lexus, Rover). In January 1994, BMW acquired an 80 per-cent share in the Rover Group through the acquisition of the British Aerospace holding (see also Colombo and Comboni 1991).

This chapter describes the importance to manufacturers of being able to understand the geographical dimensions of the market place in which they operate, and to predict the impacts of potential actions they or their competitors could take. Most manufacturers in Europe and North America have developed a capability for geographical analysis of their markets, and Section 9.3 gives a flavour of the type of capability some manufacturers now have in this area.

9.2 THE GEOGRAPHY OF NEW CAR RETAILING

There are many similarities between the retailing of cars and the retailing of any other consumer product. There is a demand which is unevenly spread across cities and regions, both in terms of volume and in relation to model type. This demand comes from two principal sources: a retail market (individuals purchasing a new or used car for their personal use) and a 'fleet market' (company cars). In the UK, the latter increased as a percentage of total sales during the Thatcher era of the 1980s, especially as tax concessions on company cars allowed employers to offer new perks to management staff. As with other forms of retailing, supply is generally provided from fixed-point locations, in this case from the 8000 or so major dealerships across the UK. However, there is a big difference in the organization of supply. The automobile industry is unique in that it has maintained a much greater control over the distribution of its products than have most other manufacturing sectors. This has been achieved through the appointment of a set of exclusive franchised dealers in virtually all countries. These dealers are not owned by the manufacturers but are independent businesses that are provided with a franchise to sell a manufacturer's product range in a defined geographical area or territory. The manufacturer agrees not to appoint other dealers within this territory in return for a commitment by the dealer not to sell any other manufacturer's products. This arrangement seems to have served the manufacturers well – they have shown little willingness to change it! The extent to which it serves the public well is a matter of some debate and a topic that attracted the attention of the Monopolies and Mergers Commission in 1992 and the European Union in 1995.

This system of selective distribution can result, especially in rural areas, in residents having little choice of which brand of vehicle they can purchase locally, if the only manufacturer represented in the area is, say, Renault or Nissan. We shall look at the implications of this for local market shares in the discussion below. Most other consumer products are not sold on this exclusive distribution basis. For example, Hitachi products are sold alongside competing electrical goods in Comet, Dixons, or Rumbelows. The motor industry responds by arguing that their products require specialist servicing and support facilities that only a dedicated dealer could supply. Interestingly, this form of retail distribution contravenes European Community competitive practice law as embodied in the Treaty of Rome. However, the auto-industry was 'block-exempted' by the EC from this specific law in 1985 for a period of ten years. Block exemption has been renewed by the EC in 1995 on a much-diluted basis for a further period of seven years. Speculation exists as to whether more power now lies in the hands of the dealers, and it is almost certain that we shall witness the growth of multi-franchising over the next decade.

As with other forms of retailing, the existence of a spatially variable demand and supply means that the study of spatial interactions becomes paramount. The magnitude and direction of these interactions will vary across manufacturer types and model types as well as between different population groups. Although the reasons for buying a new car may be a combination of personal, family, and income circumstances, there will always be an implicit geography: sports cars will sell better in more affluent areas (the influence of incomes and lifestyles) whilst four-wheel drives become more important in upland, rural areas. Readers familiar with the material in Chapter 4 will immediately note the importance therefore of

Table 9.3 Demographics and Toyota sales in the UK (*Source*: Toyota (GB) Ltd 1993)

	Age				Occupation		
	18–24	*25–34*	*35–44*	*45+*	*AB*	*C1/C2*	*DE*
Starlet	2	7	14	79	9	59	3
Corolla	2	12	21	66	19	56	1
Carina	0	7	18	75	31	46	1
Supra	2	28	38	38	63	19	0
MR2	8	44	28	19	39	45	0

Note: All figures are percentages.

geodemographics. Table 9.3 shows the breakdown of customer types for Toyota cars in the UK. It is not difficult to imagine the business planner or marketing team for Toyota purchasing ACORN or a similar geographical product to reveal areas where such population groups live. Figures 9.1a and 9.1b show the market penetration for BMW and Lada cars respectively across the UK. The geodemographics of sales clearly stand out in these extreme examples, with BMW strong in the affluent parts of south England and Lada strongest in rural northern England, Scotland, and south Wales. Similarly, Figures 9.2a, 9.2b, and 9.2c (Plates 9.1a, 9.1b, and 9.1c) illustrate the variation in market penetration (manufacturer sales/total market) by postal area for Fiat, Toyota, and Peugeot. In each case, a significant variation can be observed in market penetration by postal area. Given their respective national market shares (Fiat 2.5 per cent, Toyota 3.0 per cent, and Peugeot 6.5 per cent), there are areas of good and poor performance. For example, Toyota appears weak in Scotland but thriving in the M62 corridor and the south east. This contrasts with Peugeot, which performs well in Scotland, Wales, and the south west but disappoints in East Anglia and Northern Ireland.

The explanation for geographical variations in performance rests on a number of factors, including:

- the *variation in brand preferences* across the country. This is in part a reflection of variations in regional incomes and lifestyles. It also reflects the intensity of marketing activity – the profile of a manufacturer's customers will be identified and 'similar' types of customer targeted through direct mail. This will inherently reinforce regional biases;
- the *level of dealer representation* in an area, especially compared with competing networks.

To illustrate this latter point we draw upon two different regional case studies. The first takes the same three manufacturers and focuses the analysis on the Humberside region of eastern England. The geographical scale now

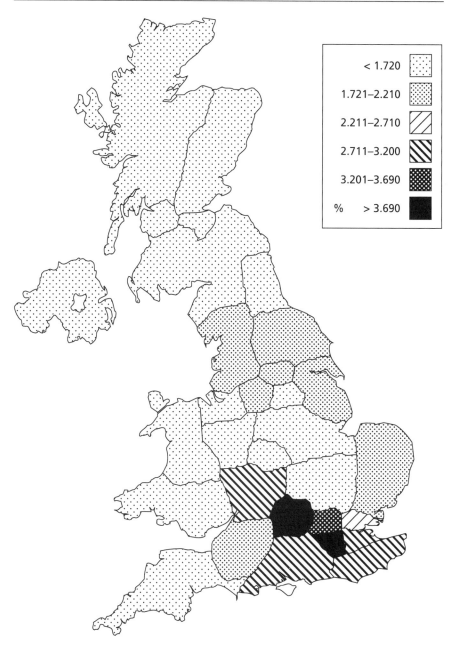

Figure 9.1a Market penetration for BMW in the UK
(*Source*: GMAP 1994)

shifts to postal districts, and the respective manufacturer's dealer points are overlaid on Figures 9.3a, 9.3b, and 9.3c (Plates 9.2a, 9.2b, and 9.2c). What

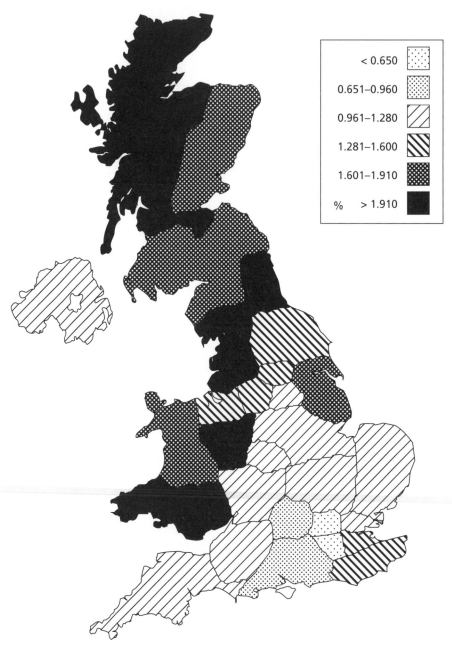

Figure 9.1b Market penetration for Lada in the UK
(*Source*: GMAP 1994)

becomes strikingly clear is that local market share is, to a large extent, driven by dealer location. For example, Fiat performs well in north

Humberside where it has six dealers (two in Hull and one each in York, Bridlington, Scarborough, and Driffield), but there is a noticeable under-performance in the Scunthorpe area where there is no local representation. The picture is rather different for Toyota, which under-performs in north Humberside (only two dealers) but has strong performance south of the Humber due, in part at least, to its relatively strong network.

The important point to come out of this simple analysis is that national market share is an aggregate of regional market share and, in turn, regional market share is an aggregate of local market share. Because of the influence of dealer location on local market share, one can recognize the importance of developing a network of dealers that are in some way optimal with respect to the market and also to the location of competing dealers. This suggests that manufacturers should have a very dense network of dealers to maximize local market share. However, as discussed earlier, motor dealers are independent businesses in their own right and rely upon achieving reasonable levels of sales to maintain their viability. Too many dealers would lead to low average sales per dealer and lack of financial viability. For manufacturers there is therefore a trade-off between maximizing market share through a relatively large dealer network and maintaining dealer viability through reducing the number of dealers in direct competition. Solving this problem is not easy, but it is the province of spatial modelling and, as we hope to demonstrate later in this section, an area where considerable business benefits can be generated through the use of SDSS or IGIS.

The second example deals with Toyota and its performance in north Wales. In 1987 Toyota sold fewer than 100 cars in the whole of north Wales, producing a very low market share of 1.6 per cent (see Figure 9.4). What explanation could be offered for this? Perhaps the demographics were simply wrong – not the required level of incomes perhaps. Yet successive volumes of *Regional Trends* (the HMSO annual publication) indicate that the levels of affluence in north Wales are comparable with those of other semi-rural areas such as Devon and Cornwall. Here, however, the Toyota market share in the same year was 2.2 per cent. A more obvious solution lies in the fact that the only dealerships in north Wales were at Harlech (closer in fact to mid-Wales) and Wrexham (to the far east), with a neighbouring dealer-ship across the border in Chester. Faced with so many intervening opportunities, consumers are unlikely to be prepared to search so far from home for a new car, or to be prepared to travel such inconvenient distances for servicing. There are obvious individual exceptions to this, but these are few. With the introduction of a new dealership in Llandudno, Toyota was able to increase its market share in the Conway Valley areas to around 10 per cent by 1992.

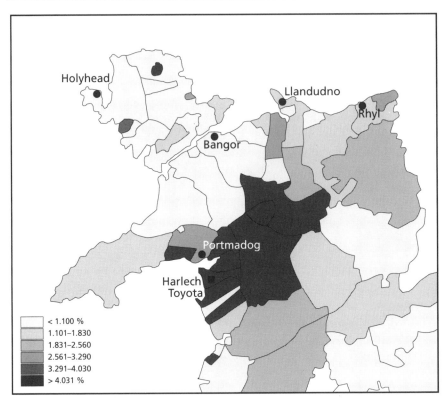

Figure 9.4 Market penetration for Toyota in north Wales, late 1980s
(*Source*: GMAP 1994)

The relationship between market share and the dealer network is fundamental to the remaining arguments in this chapter. Table 9.4 shows that at
the national level there is a dense network of dealerships, producing variations in average sales per dealer from 1100 for Ford to only 180 for Volvo
and VAG. Clearly, building more dealerships will produce more sales; but
there are diminishing returns as the network becomes saturated, and the
skill is to plan geographical location strategies more effectively. Vauxhall,
for example, are able to generate a market share of 16 per cent from just 601
dealers, an average sales performance of 580 per dealer. The quality of the
product and sales staff are, of course, also very important, but as product
quality and price are likely to converge further during the late 1990s (at
least by model segment), the battle for market share will increasingly be
from the showroom, with all the implications this has for getting those
showrooms in the right locations.

Given the above discussion, it is not surprising that the car industry has
increasingly investigated products available to aid geographical market

**Table 9.4 Average sales per UK dealer
1994 (*Source*: GMAP 1994)**

Brand	Sales
Ford	1100
Vauxhall	580
Rover	320
Peugeot	400
Renault	340
Nissan	340
VAG	180
Citroen	320
Fiat	350
BMW	380
Toyota	200
Honda	240
Volvo	180

appraisal and dealer planning. Apart from geodemographic systems (see Chapter 4 for more details), manufacturers have also looked to GIS. Clifford (1993) shows how Vauxhall have invested in a GIS principally to aid the understanding of local performance. This is achieved by mapping sales in relation to dealers and their assigned sales territories. Figure 9.5 shows a good illustration of this. In addition, sales can be overlaid with demographic information to try and understand the relationship between models sales and population types. Clifford (1993) goes on to describe the early euphoria associated with implementation, but some subsequent reservations as the system became operational.

In order to understand the problems with this sort of analysis, it is necessary to recall much of the discussion in Chapter 3: namely, the lack of any real analysis of the workings of the car market that is possible using this type of GIS, and the obsession with relating sales to drive time bands (or the circular catchment area problem we identified in Chapter 3) rather than any real treatment of spatial interactions and the complicated influences of competitor location. For these reasons, and other reasons such as ease of use (Clifford 1993), such generic systems are not likely to be the most successful. Indeed, it is our experience that a growing number of executives within the car industry are now looking to decision-support systems for a more complete solution. The next section describes the bases of an IGIS designed for two large motor clients, Toyota and Ford. We then continue to examine the use of the system and the important geographical questions they aim to address.

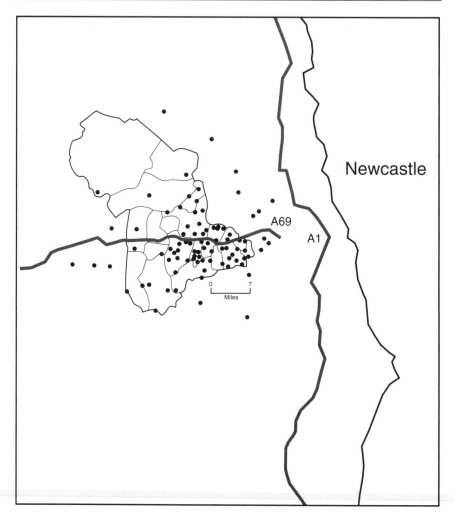

Figure 9.5 Use of GIS by Vauxhall for sales analysis
(*Source*: adapted from Clifford 1993)

9.3 THE DESIGN OF A SDSS FOR THE MOTOR INDUSTRY

9.3.1 The background

The remainder of this chapter draws upon GMAP's experiences with two major manufacturers, Toyota and Ford. Our relationship with Toyota began in 1987, with a report on trading performances in two pilot regions of Oxford and Merseyside. Following the success of these studies, Toyota commissioned a full UK national modelling and information system for new car registrations. Work with Toyota has subsequently been extended to include an information and modelling system for parts and servicing in the UK, a further new car registration system for Toyota Belgium, and a separate investigation into the optimal locations for retailing a new top-of-the-range model (Lexus). Our relationship with Ford began in 1989, when GMAP successfully bid for the development of a spatial decision support system to assist all areas of Ford's distribution departments: market representation, marketing, sales planning, and more local district management. The subsequent system, RADAR (Retail Analysis of Dealer Areas of Responsibility), is directed at the main market segments of new car sales, parts and servicing, and commercial vehicles. The history and development of RADAR are discussed in some detail by Whitton (1993).

Prior to working with GMAP, both companies undertook 'manual' market studies of various UK regions, involving periods of intensive fieldwork. Typically it would only be possible to study two or three regions in depth in any one year, because of the time involved in putting all the data together on the car market and on geodemographics, particularly as most of the data existed in paper form. Hence, any one region (say Merseyside) may only have been fully analysed once every 10 to 15 years. Alongside the data collection process, market representation staff would 'walk the streets' to appraise the local situation in greater detail. Such site visits remain crucially important to support the data collection process, but it is now possible to handle the latter in a number of weeks rather than three or four months, because the decision support system can highlight which parts of the system warrant attention. In addition to data collection, however, such information systems offer new methods for analysing those data. Previously it was very difficult to combine the data sets in any meaningful way, other than relying on gut feeling for an appreciation of the apparent variations. As outlined in Chapter 3, models within GIS environments offer considerable support for data combination, linkage, and analysis. We shall describe the construction and use of such systems below.

9.3.2 Building an information system

Unlike many retail sectors, the motor industry has excellent information relating to sales by small geographical area. Not only do manufacturers obtain detailed sales information on their own dealers, but they can also capture data relating to competitor performance. In the UK these sales data are recorded at the Driver and Vehicle Licensing Centre (DVLC) at Swansea whenever a new car is purchased within the UK. Every dealer selling a new vehicle has to complete a form (V55) which identifies the precise specification of the vehicle and details concerning the purchaser. In turn, these data are made available to the Society of Motor Manufacturers and Traders (SMMT), which releases them to members (such as manufacturers) on a syndicated basis. For individual manufacturers the following information is available:

- new car registrations for each manufacturer by model type for every postal district in the UK;
- for a particular manufacturer or distributor, the postal district of each new car registration by model type and by selling dealer code;
- for each manufacturer, the total number of registered cars by model and year of first registration on the road by postal sector (CAR PARC data).

Combining these three data sets provides the manufacturers with a unique information base relating to both total sales and the spatial interactions between demand and supply. If we select one manufacturer alone (say Toyota) and analyse data at the postal district level, then for the nine major model segments they compete in (Corolla, Carina, Camry, MR2, etc.) we have interaction arrays of enormous size:

$$2800 \text{ postal districts} \times 9 \text{ model segments} \times 200 \text{ dealers}$$
$$= \text{interaction array of } 5\,040\,000 \text{ elements.}$$

If the 25 main manufacturers are added into the equation, one can begin to appreciate the difficulties of using any sort of manual (non-computerized) system to handle the complexities of the geographical markets. The benefits of a GIS just to store, retrieve, and map any of the elements of the arrays are significant. If one adds time series data, the possible number of spatial queries becomes enormous. From the main SMMT data sets, the following sorts of spatial query (in map or tabular form) become possible:

- sales by manufacturer and model by small area;
- sales by dealer by model;
- change in sales by manufacturer by model by year;
- change in sales by dealer by model by year.

In the United States, similar types of registration data are available, collected on a state-by-state basis. Here, GMAP works with the RL Polk company, which collates the state registrations and produces a national database covering all movements, which is marketed to the various manufacturers and importers. Most European countries have similar arrangements for the dissemination of registration data on a syndicated basis.

GMAP and RL Polk have collaborated in the development of an IGIS for the US auto industry, which is marketed under the generic title of Decision Point. The illustrations of Decision Point presented here are all taken from the marketing prototype, which has been developed using 1990 data for the Seattle/Tacoma study area. Individual clients of Decision Point have been supplied with customized systems to meet their own requirements, and therefore the following examples should be seen as illustrative of the general approach but do not necessarily represent the approach of any particular manufacturer.

The Seattle/Tacoma region in Washington State is located on the Pacific seaboard. The geography of the market is complex – a series of islands, lakes, archipelagos and bays make communications difficult. There were, in 1990, a total of 108 658 new vehicle sales generated from 192 dealerships. The Asian manufacturers are particularly strong on the West Coast of the USA, accounting for nearly 50 per cent of all registrations. To demonstrate that the same type of relationship that exists between dealer location and market share in Europe is also present in North America, we can refer to Figure 9.6a (Plate 9.3a). This shows Ford market share and their dealer locations for the Seattle/Tacoma region. Broadly, there is a strong correlation between peaks in market share and dealer locations, although the relationship sometimes breaks down, for reasons that are explained when we investigate the data further: a poor performing dealer or a large fleet purchaser may well distort the figures.

We can now present some examples that illustrate a range of the features of the information system component of Decision Point. Figure 9.6b (Plate 9.3b) compares the pattern of sales of an Audi dealer in eastern Seattle with that dealer's ability to penetrate the Audi market in Seattle, as measured by the percentage of all Audi buyers in each Census tract that buy from that dealer. We can observe that in some Census tracts close to the dealer, less than 40 per cent of Audi buyers buy from their local dealer. This is a clear demonstration that customers do not necessarily frequent their nearest dealer. It does remain an interesting question for this particular dealer as to why consumers would choose an alternative dealer further away.

One explanation for this apparent irrationality is that consumers may wish to purchase a new car from a dealer near where they work rather than near where they live. Figure 9.6c (Plate 9.3c) compares the pattern of sales within the Seattle area of two dealers – one in the city centre, one in the suburbs. The suburban dealer generates high levels of sales in the vicinity of the dealership, with the expected distance-decay effect beyond that. The downtown dealer has a much more dispersed pattern and achieves sales in areas that are considerable distances from the dealership. Indeed, the highest level of sales for this dealer are achieved in Census tracts nearer the suburban dealer than his own location. Our conclusion from this type of empirical evidence is that manufacturers should be wary of the more simplistic, brochure-based analysis that often characterizes traditional approaches to dealer planning. Consumer behaviour is complex, but it is still predictable using the right kind of spatial interaction tools, which we have been advocating in this book.

Figure 9.7a (Plate 9.4a) illustrates how simple information can be combined to present a visual summary of the potential opportunity in the Seattle market for Pontiac (a division of General Motors). This map has been generated by comparing the actual market share the manufacturer achieves in a Census tract with the average for the Seattle market. If the former is smaller than the latter, we calculate the number of additional sales that would be gained if average market share was achieved. We can see that in the resultant map there is a strong concentration of opportunity in the eastern suburban area. We can also note that Pontiac does not have a dealer in the area (the red squares are Pontiac dealers in this market). We could now use the predictive modelling capability of Decision Point to test the impact of opening a new dealer somewhere in this area. We perform this analysis later in the chapter.

Manufacturers are often interested in the relationship between the socio-economic and demographic characteristics of the population and those of their customers. Understanding this relationship can help manufacturers and individual dealers target potential customers in certain areas. Figure 9.7b (Plate 9.4b) identifies the number of households with greater than $60 000 income (1990) in each Census tract, and compares this with the pattern of sales generated by an Audi dealer in east Seattle. Audi would be regarded as an executive brand in the US and should therefore appeal to high-income households.

It can be seen that the dealer is located within a high-income area and generates good sales volume in tracts with large numbers of high-income households. However, there are additional Census tracts to the north and south of the dealer where no sales are achieved but which have significant numbers of high-income households. The manufacturers and/or the dealer could use various marketing devices (direct mail, poster advertising) to attempt to increase sales penetration in these areas. This local intelligence can be used to generate marginal improvements for each dealer. Replicated across the country (there are 90 or so markets of a similar size to Seattle in the US), it could add up to a useful overall improvement in sales for the manufacturer.

These examples give a flavour of the types of information that can be generated from the first component of Decision Point. The main use of this sort of information is to produce systematic and regular market studies of individual dealers and their catchment area, or broader studies relating to cities or regions. Clearly, the analyst can examine where sales or market gaps may exist (a precursor to looking at strategy responses in the next section). The same analysis can be undertaken for the competition in the same localities.

The information system has obvious advantages for marketing and advertising, and, where time series data are available, for monitoring the impacts of advertising or marketing campaigns. If sales are low in a region, can the dealer target those areas for local media advertising and ascertain whether this brings any response in the long term? Whilst marketing is a complex activity based on increasing product awareness, there are still evident gains in profiling the geodemographics against dealer performance.

9.3.3 Adding a modelling component

Previous chapters have argued that the construction of the information system is only the first half of the strategy for a complete IGIS or SDSS. In the motor industry, we are equally concerned with all the 'what-if' questions outlined in Chapter 3. The next section will discuss these in detail. To complete a SDSS such as Decision Point, a basic spatial interaction model appropriate to the car market needs to be outlined:

$$S_{ij}^{mk} = A_i^{mk} O_i^{mk} W_j^{mk} f(c_{ij}^{-\beta k}) \qquad (9.1)$$

where:

S_{ij}^{mk} = registrations in postal district (i) attributed to dealer (j) for manufacturer (m) and model segment (k) ;

O_i^{mk} = registrations in postal district for manufacturer and model segment;

W_j^{mk} = attractiveness of dealers in terms of facilities provided at the dealer site, such as size of facility, showroom floor spaces, etc.;

c_{ij} = travel between postal district and dealer;

β^k = parameter to measure the friction of distance for model segment k;

A_i^{mk} = balancing factor to ensure:

$$\sum_j S_{ij}^{mk} = O_i^{mk}$$

This last constraint implies that we have a fixed market size by small area and model segment. New dealers are not assumed to generate new demand *per se*, but merely to facilitate exchange or switching between dealers and brands.

Although this basic model is indicative of the type of formulations used in GMAP's commercial applications, it is a simplification of actual models implemented and calibrated. To achieve true realism a number of important factors need to be added. Specifically, these relate to:

- relative brand competitiveness by manufacturer and model segment. We know that different manufacturers have different brand strengths that relate to quality of product, image, and reputation. In the UK, for example, Ford are much stronger than Fiat, but the situation is reversed in Italy. However, Ford in the UK are relatively weak in the executive market. These factors must be incorporated into the models to reflect dealer attractiveness accurately;
- consumer preferences and loyalties for one brand over another.

(See Chapter 5 for more details on model disaggregation.)

Having described the model, its many potential uses in a variety of different planning situations can be explored.

The analysis of the information system will have given the market planner an excellent picture of the performance of his or her company in relation to its competitors in whichever geographical region. That planner is now faced with putting together an action plan of some kind in response. The first issue relates to judging existing dealer performance. The maps produced by the information system help to describe the spatial pattern and show obvious areas of poor/good performance in relation to total sales and market share. However, it is still difficult to decide objectively whether that performance level could realistically be expected to be higher (given the levels of demand and of competition). It is obvious that 100 sales per year is preferable to 50 sales per year, but should that dealership be capable of selling 120 or 150 cars per year? Analysing market penetration figures gives a better indication of performance, since this allows the analyst to determine how many zones surrounding the dealership achieve higher than national average market penetration figures. However, it is possible to devise a new index of performance using the baseline model calibrations. That is, given demand levels, the nature and volume of the competition, and the average drive times to dealers, what do the models predict the sales of a particular dealership *ought* to be if it were an average national performing dealership? Such predicted revenue totals are easily calculated by the models. This is a useful measure of performance, since it is calculating the market potential of an area rather than relying solely on actual sales achieved.

Having assessed existing dealer performance across a city or region, the company may now wish to explore a variety of scenarios for change. If performance is judged to be uniformly good, it is likely that the existing network of dealerships is serving the area well. In this case, the future strategy is one of securing and maintaining market penetration levels in the face of the strategies of competing retailers. It is here that marketing and advertising have crucial roles to play. Indeed, it is normally the case that companies spend more of their advertising budgets in regions of high performance in order to protect their market shares (Clarke and Bradford 1989).

If performance is deemed to be below acceptable levels, more radical action may be required. There are a number of possible scenarios here. First, the information system may have thrown up some clear gaps in the network, which appear to be sizeable markets in their own right. These are possible opportunities for new dealer locations. As shown in the previous chapters, the models can simulate the opening of new dealerships in these locations. By adding a new dealership, both the likely revenues, in terms of predicted cars sold, and the deflections from the company's own existing dealers and those belonging to the competition can be estimated.

Reverting to our Seattle example, we recall that we identified an area in the east of the city where there appeared to be good potential for a Pontiac dealership. The next series of maps and tables illustrates how Decision Point is used to record the impact of opening a new dealership in a Census tract in this region. Not only will the models estimate potential sales, they will also predict revenue totals by model segment (superscript k in equation 9.1). This would allow the dealer stock inventory to be more carefully monitored and organized for local market conditions. This is crucially important, given the expense of retailing different models from different outlets.

A second possible conclusion to the market appraisal is that there are currently enough dealers (that is, market coverage is generally good), but a number of these seem to be performing poorly. The reasons for this may be numerous, including poor local site conditions, poor management, poor salespersons, or very good performances from competing dealers. The solutions to this problem will thus vary according to which of these circumstances actually prevail. If the problem lies with the site or staff of the existing dealership, dealer termination or relocation may be a possible option. Again, the full impacts of dealer termination/relocation can be assessed by feeding the changes into the model and calculating the impacts. The particular example we have chosen from Decision Point to illustrate how this process works in practice is the relocation of a dealership. In the US, General Motors operates under four main brands: Chevrolet, Oldsmobile, Pontiac, and Cadillac. There has been much press speculation that GM will merge two of these brands into one, although at the time of writing this has not occurred. For our example, we note that there are both an Oldsmobile and a Chevrolet dealer in close proximity in south-east Seattle. Figure 9.8a (Plate 9.5a) effectively relocates the Oldsmobile dealer to the Chevrolet dealer site. Figure 9.8b (Plate 9.5b) then compares the total sales from the proposed relocation. We can see that the total Oldsmobile sales decline from 1045 to 935, but there may be considerable efficiency savings by operating from a single dealer point rather than two that offset the reduction in profitability from the decline in sales. Figure 9.8c (Plate 9.5c) shows the gain in sales from nearby Oldsmobile dealers, which clearly benefit from the closure.

A third conclusion possible from the market appraisal is that, whilst the dealers are achieving reasonable local market shares, they seem to be in generally poor locations in respect to the competition. This may be purely the result of historical legacy, with the dealer network being built up over a long period of time when market conditions might have been very different from today. In the short term, the solutions may involve some combinations of the above strategies of dealer relocation, termination, and new appointments. In the longer term, the company may be interested to know what the optimal locations for its local network currently are, given the objectives of maximizing either total sales or market shares. The constraints on such an optimization procedure might include an index of dealer viability (no dealer must sell fewer than 100 cars, for example); a maximum or minimum number of dealers; the inclusion of certain existing dealer locations; a minimum drive time between dealers (ensuring a more uniform geographical spread of dealers); and the exclusion of 'non-feasible' sites, which might include those unlikely to gain planning approval. Formally, the spatial interaction model can be rewritten as a mathematical programming formulation with the following objective and constraints:

Maximize: Market share in region X for manufacturer Y

Subject to: Maximum number of dealers
Minimum number of dealers
Minimum dealer sales of Y

Minimum inter-dealer drive time of T minutes
No consumer to be more than M minutes' drive time from a dealer

Essentially, the backcloth of other dealer locations is kept fixed and demand estimates for the current time period or some projections for future years are used. A heuristic algorithm has been developed that solves this complex problem on a PC (for a full description of the detail see Birkin et al 1995).

To illustrate the results from this approach, we return to the analysis undertaken with Decision Point in the Seattle market. Toyota have a current representation pattern in Seattle, as shown in Figure 9.9a (Plate 9.6a) (note that the blue lettering refers to current dealers). The task was to use the model above to ascertain whether any new opportunities for Toyota dealerships existed within the Seattle market. The criteria used for the optimization model were as follows: there must be a minimum and a maximum of two dealers in the region: no new dealer allowed within 13 minutes' drive time of an existing Toyota dealer; each new dealer to have a minimum of 300 new car sales per year; no potential customer to be more than 25 minutes away from a Toyota dealer.

The results of this analysis are presented in Figure 9.9a (Plate 9.6a), which highlights the location of the two new dealers: one in the downtown area and an additional dealer in the prosperous eastern suburbs (here the red squares identify the two new dealer locations). In Figure 9.9b (Plate 9.6b), we can see that the two new dealers sell 576 and 492 new cars per year respectively. In total this 1068 volume sales through the new dealers equates to 850 incremental sales increase for Toyota, as some of the sales generated at the new locations are deflections of sales from existing dealers. Simple arithmetic suggests that if this type of incremental improvement could be obtained in each of the 90 metropolitan markets in the USA, Toyota could generate more than 75 000 additional retail sales per annum. The impact on profitability would be significant (see Chapter 11, and Clarke and Clarke 1995).

A final example of the use of this optimization procedure examines a 'clean-sheet' scenario. Suppose we could develop our dealer network in a market from scratch: where would our dealers be compared to the current situation? To illustrate this we use Lincoln (the luxury division of Ford) as an example. They currently have eight dealers in Seattle selling a total of 538 cars per year (the red squares in Figure 9.10a (Plate 9.7a)). Suppose we asked the model to identify an optimal network where no dealer sells fewer than 67 cars per annum and all Lincoln dealers are at least 20 minutes' drive time from each other (see the scenario editor in Figure 9.10b (Plate 9.7b)). The results of this are shown in Figure 9.10a (Plate 9.7a): six dealers are identified which generate 466 sales per year. (The heuristic has actually identified two dealers with slightly fewer than 67 sales, which is a characteristic of the methodology which we can ignore for the purposes of this discussion.) This is only 72 sales fewer than the current situation, but with 25 per-cent fewer dealers. We can see some correspondence between the two patterns, except that the model identifies only one dealer in downtown Seattle (44) and one in Tacoma (715.01) compared to the two existing dealerships in these areas. However, we could easily change the scenario constraints and review the model to see how things change.

9.3.4 Summary

Whatever scenarios are tested, the important point to stress is that the SDSS or IGIS allows proactive planning. Thus it is possible to use the system intelligently to formulate a local or regional policy and to test that policy in a controlled environment. However, it should not be forgotten that all the routine reactive questions can also be assessed and answered using the system. These might include enquiries from potential new dealers wishing to transfer allegiances to a new company (hence allowing the system to be used in the same way as outlined above) or from existing dealers wishing to transfer elsewhere. Similarly, competitor changes (such as the opening of a new dealership) can be monitored and evaluated in the same way: what is the impact of a new Ford dealership likely to be for Toyota sales? Where would Ford gain sales and at whose expense? How can we, Toyota, react to minimize the damage of such competitor behaviour?

In addition to generating local market plans for dealer developments, such a SDSS or IGIS can be used in a more *ad hoc* way to answer other specific problems the company may have. For example, the SDSS or IGIS can help launch a new product, such as a luxury sports car. The marketing team will have done their homework to predict the likely customer target group. The major strategic question facing the dealer development department is which franchises should be considered for retailing the new car. These would have to be dealers within the network that would be able to achieve a minimum level of sales to warrant the award of the new franchise.

9.4 CONCLUDING COMMENTS

Since the installation of the first versions of the SDSS for Toyota and Ford, most of the above scenarios have been addressed and evaluated by the companies themselves. Most importantly, a blueprint has now been developed for the future expansion of the respective franchised networks, and both organizations will not undertake future network changes without putting the scenarios through the SDSS. Whitton (1993) gives some illustrations of both the proactive and reactive uses of the Ford RADAR system during a typical two-week period in July 1992. On a reactive basis, the market representation team were asked to use RADAR to look at the performance of an existing dealer serving east Yorkshire and Humberside, and to identify and evaluate potential new sites for new dealerships in the Reigate, Dorking, and Woking areas, plus Birmingham and London Docklands. In addition, there was a potential relocation of a Ford dealer in Crawley to investigate. On a more proactive basis, Ford recognized that the whole of the Greater Manchester area was producing only 17 per-cent market share, well below performance levels elsewhere. In only 14 days or so, it was possible to produce a full report on the best strategies for expansion, as outlined in Section 9.3.

One of the great synergies that exist between the auto industry and GIS applications is the strength of the underlying registration database that exists in most developed countries. This incorporates both small-area detail and model sector disaggregation. More than this, however, the fact remains that the auto industry has recognized the fundamental geographical dimension to its business. It has done this more so than any other industry sector. It is also a truly global business – GM and Ford are two of the largest corporations in the world, with annual operating revenues greater than $100 billion. Given this scale, it is clear that small incremental advantages at the local level repeated systematically across regions, nations, and continents can generate large returns. IGIS has already begun to inform this process through the better understanding of geographical markets and, as demonstrated in this chapter, through the optimization of distribution networks. There are many more ways in which spatial analysis can contribute to the effective planning of a manufacturer's business in areas such as sales planning, marketing, and the 'after-market' business. However, the biggest opportunity might lie in the contribution to strategic as well as tactical issues in the auto industry.

There is massive structural change taking place on the manufacturing and distribution sides of the industry. This involves huge capital investment in facilities, equipment, logistics, and distribution. Geographical analysis has a crucial role to play in understanding and planning the entire 'manufacture-to-market' process. It will be interesting to observe how much impact IGIS has in assisting this challenge.

CHAPTER 10

Spatial analysis and the geodemographics of financial services

10.1 INTRODUCTION

The financial service sector is at the forefront of economic growth and change in the UK and many other developed countries. This state has been fuelled by a variety of factors, but little research has been undertaken to evaluate the methods by which this growth could be managed or optimized. Retail financial services, in particular, have a variety of features that have an important spatial dimension, and as such it would be expected that the range of spatial analytic methods that have successfully been used in retailing and in other service industries would have found their way into the financial services sector. However, their use in this sector remains in its infancy. The major aim of this chapter is to illustrate how both traditional and new methods of spatial analysis can be developed and applied in the planning and management of financial services institutions.

One of the main reasons why these methods have largely been ignored is that, until recently in the UK, the level of competition between the main retail financial service institutions such as banks and building societies was very low. For instance, prior to 1971, there were cartel agreements to set interest rates and hence limit competition. The 1980s saw considerable deregulation of the financial service industry, a consequent influx of US, European, and Japanese banking institutions into the UK market, and the widening of the services provided by financial institutions: for instance, banks providing mortgage funds and building societies providing unsecured loans. The Stock Exchange 'Big Bang' of 1986 has revolutionized the trading of stocks and securities, and the interest in, and ownership of, shares has been encouraged through the large-scale privatization of many state-owned utilities and companies. The net effect of all this activity is a significant increase in the level of competition between institutions. This in itself has forced institutions into cutting costs and increasing the effectiveness of their marketing. Some of the consequences of these trends are examined in Section 10.2.

Financial service retailing is essentially different from traditional retailing for a number of reasons. First, the decision to purchase (become a current account holder or credit-card holder, buy a pension plan, acquire a mortgage, etc.) is made very infrequently – in many cases, only once in many years. Secondly, there is often confusion of purpose. When purchasing a fridge or a loaf of bread, a basic need or demand is being satisfied. The same cannot be said of opening a current account, where the needs are varied. Thirdly, there is no clear product differentiation. Many services are substitutable and interlinked, such as mortgages with life assurance and/or different types of extra-interest deposit account. Fourthly, many financial services are of relatively high value, and a good deal of trust is required between consumer and supplier. When an individual deposits £10 000 in an account or purchases a pension plan, he or she expects to be able at least to get his or her money back and draw a pension respectively. The final difference, which may be relatively minor, rests in the fact that services are frequently acquired from a person's workplace in addition to her or his home location.

Despite these differences, however, there are a number of obvious similarities, most notably the requirement for matching demand and supply over space. As we shall see, there is plenty of scope for adapting methods developed to assist this process in the traditional retailing environment for use in the financial services sector.

The rest of this chapter is organized as follows. Section 10.2 describes the growth and changes in the UK financial services market. This has been both supply- and demand-led, with institutions developing new products and consumers becoming more sophisticated and discerning. Section 10.3 examines systems of spatial organization in financial services. Here we distinguish between two different but related types of service: those that are mainly transaction-based (such as cheques, credit cards, ATMs, etc.) and those that are mainly non-transaction-based (such as loans, mortgages, insurance, pensions, etc.). Some of the main spatial problems faced by institutions in developing these service are described. Section 10.4 examines how financial service institutions identify and attempt to find their customers. In particular, it describes and evaluates methods such as geodemographics for use in marketing and the link with proprietary GIS. Section 10.5 illustrates how spatial analysis could be applied to help in three typical financial application areas: branch evaluation and rationalization, television advertising optimization, and direct mailing. This section also outlines the framework for an integrated approach to financial service marketing and planning in general. Section 10.6 summarizes the main conclusions of the chapter and suggests directions for future development. Section 10.7 offers concluding comments.

10.2 THE GROWTH OF FINANCIAL SERVICES

Recent years have seen an enormous growth in the purchase and usage of financial services by individuals. The period of the mid-1980s saw the most rapid growth of financial products. In the period 1982–86, for instance, deposits with banks and building societies increased by 123 per cent and 15 per cent respectively (CSO 1987) and borrowing by the consumer sector increased by 65 per cent (CSO 1987). Although the volume of transactions peaked around 1990, the savings ratio increased from 5.7 to 12.6 between 1988 and 1992. This growth has been both demand- and supply-led, with, on the one hand, increasingly discerning, product-hungry customers and, on the other hand, an increase in the number of market competitors in each financial product area.

This upsurge was encouraged by an increase in the total personal disposable income by over 35 per cent during the period 1982 to 1986 (CSO 1987), by increased levels of financial education and awareness, and by government legislation, such as the income redistribution caused by taxation policy, the broadening of the nation's asset base by actions such as the sale of council houses, and the broadly based privatization of nationalized industries.

Market competition has been encouraged by a series of legislative changes within the UK market place and by reductions in the rules and regulations covering financial trading. This started with the 1971, government paper 'Competition and Credit Control', which abolished the banks' interest rate cartels, promoted free banking competition, and allowed foreign banks entrance to the UK market. Since 1971, all forms of financial institution have been diversifying widely and encroaching upon the markets traditionally associated with other financial institutions. The banks no longer monopolize transmission facilities, the building societies savings and mortgages, and so on.

The boundaries between the various financial institutions are weakening and are likely to do so further, as represented in Figure 10.1. The banks, for instance, are now heavily committed to home mortgage provision – traditionally the province of the building societies. Banks and building societies have both entered the estate agency market, and, with the recent deregulation of building societies, they are able to offer money transmission services and personal loans, both secured and unsecured – traditionally the province of the banks and finance houses. In more recent times, banks and building societies have developed a wide range of insurance and pension services. In the first instance, this was with partner organizations (such as the Halifax and Sun Alliance), but more commonly they are now developing their own 'bank assurance' products internally (for instance, NatWest Life). The increasing level of market competition has led the institutions involved to adapt their traditional business techniques to narrower profit margins.

This in turn has resulted in:

- the acceptance of economies in resourcing;
- a search for new and cheaper distribution channels, such as telebanking;
- competition based upon service levels and quality;
- the wider adoption of traditional consumer marketing techniques.

The first item speaks for itself, with staff cuts, systems streamlining, and the substitution of staff by technology apparent within all financial institutions.

Additionally, and more importantly from a geographer's viewpoint, it has led to experiments with the service distribution channels. All banks and many building societies, for example, have installed ATMs to reduce staff costs in branches. Most institutions are now actively involved with EFTPOS (electronic fund transfer at point of sale) technology, some have closed branches, and others have experimented with the functional responsibilities of the branches, creating inter-branch hierarchies to service customers more economically.

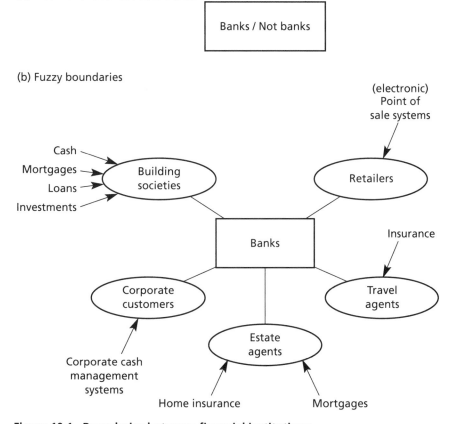

(a) Traditional clear-cut boundaries

Banks / Not banks

(b) Fuzzy boundaries

(electronic) Point of sale systems

Cash
Mortgages
Loans
Investments

Building societies

Retailers

Insurance

Banks

Corporate customers

Travel agents

Estate agents

Corporate cash management systems

Home insurance

Mortgages

Figure 10.1 Boundaries between financial institutions

The quality and breadth of service available from individual institutions has become an important competitive standpoint, with longer opening hours, personal advisers, and a friendly welcome becoming key features of many successful marketing campaigns.

In conjunction with 'service quality', competition, and marked differentiation techniques has come a reliance on some of the more traditional consumer marketing practices, such as price cutting, new product proliferation, free 'give-aways', and direct sales and advertising techniques. Examples of the importance of price cutting can be seen in almost any financial medium today – cheap mortgages, cheap loans, no survey fees or valuation estimates for mortgage acceptance, and so on. Free 'give-aways' in the financial sphere include the prolific children's handouts available from all the big banks, American Express rewards for introducing new members, cash back on mortgages, car companies entering the credit-card market, and more recently customer incentives to use particular cards. However, with the housing market being relatively depressed in the mid-1990s, we have witnessed cash backs and even free cars being offered to tempt customers back into the market place.

The importance of direct sales and advertising techniques is increasing dramatically within the financial sphere: no longer is the branch premises the only venue for product/service sales, but now individuals, both customers and non-customers, may expect to be targeted whilst watching the television, listening to the radio, answering the telephone, opening their mail, reading the newspaper, or even driving down the road looking at the various poster sites. In 1990, £110 million was spent by the clearing banks alone on advertising – a dramatic increase from almost £18 million in 1980.

To summarize, the competition for financial service provision has increased dramatically over recent years, and financial service providers are looking increasingly to retail market techniques to help them compete. The geodemographic surveys and spatial analyses of demand and supply used in retail analyses could have a major impact upon the success of all these new marketing initiatives for financial institutions, but as yet they are not being used to their maximum potential. Examples of how spatial techniques could improve performance in three key areas are given in Section 10.5.

10.3 ISSUES OF SPATIAL ORGANIZATION IN FINANCIAL SERVICES

10.3.1 Introduction

In the face of the increasing proliferation of products sold by each financial institution, and the inability of branches to sell and service the new products (a method which would anyway not be cost-effective), institutions are looking increasingly to direct sales and remote servicing techniques for their products. As a consequence there is an increasing polarization of products into two categories:

1 transaction-related;
2 non-transaction-related.

Each requires a different consideration in terms of its cost-effective distribution. Transaction-related products are those requiring frequent interaction with the supplier, either directly over a branch counter (for instance, current accounts) or indirectly via a retailer (for example, credit cards). Non-transaction-related services are those where, apart from the initial purchase decision, no customer contact is undertaken (for example, mortgages and pensions) and additionally no detailed servicing infrastructure is required.

In each case, there are two distinct spatial issues: those related to sales and those related to delivery. Table 10.1 illustrates the kinds of issue that are faced in each of the two types of service and the two areas of concern. All these elements have an implicit spatial component. In attracting and finding potential customers, much use is made of geodemographic targeting (see Section 10.4). In cutting costs and improving the effectiveness of the delivery mechanism, attention has to be focused on optimizing the branch network structure, availability of ATMs, and so on. The rest of this section examines the spatial aspects of transaction-based services, while Section 10.4 examines the issues involved in geodemographic targeting.

10.3.2 Transactional-based services

Under this heading are considered cheques, cash withdrawals from ATMs, credit- or charge-card transactions, and the recently developed Electronic Fund Transfer at the Point of Sale (EFTPOS).

Table 10.1 Issues related to sales and delivery

Mechanism	Transaction	Non-transaction
Sales	How to attract and find customers	How to attract and find customers
	How to make customers continually use the product	
Delivery	How to cut costs regularly	How to cut costs
	How to improve customer service	How to provide service

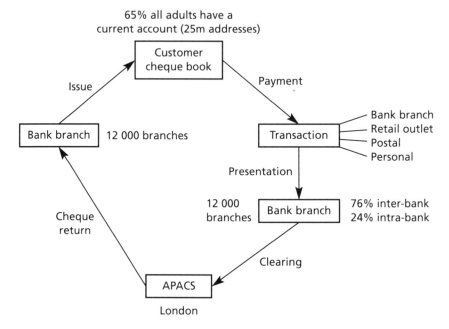

APACS = Association for Payment Clearing Services

Figure 10.2 Cheque-clearing process

10.3.2.1 Cheque books and current accounts

Although used in the UK for over 250 years, current accounts, and hence cheque books, have only been held by significant numbers since the 1950s. In 1976, 45 per cent of adults had a current account, and this had risen to 65 per cent by 1985 and 77 per cent by 1991 (Burton 1994).

The process of cheque issue, payment, and credit/debit is quite complicated in terms of both the number of parties involved and the amount of paperwork. Figure 10.2 illustrates the main processes. A bank branch issues a cheque book to an account holder on request, usually by post. The account holder is then free to use cheques for payment, either by post or by person in bank branches, retail outlets, and so on. This process is normally accompanied by a cheque guarantee card (in use since 1964), which guarantees payment to the recipient of the cheque, up to a maximum amount. The recipient of the cheque then presents it at his or her own branch (which will usually be both a different bank and in a different location from that of the issuer). A central clearing service in London, known as the Association for Payment Clearing Services (APACS), is used by all clearing banks for the organization of the fund transfer between account holders.

The cheque itself is returned, cleared, to the issuer's bank branch. In 1992, some 2.4 billion cheques were cleared in this way – approximately

10 million per working day. Being paper-based, the system is enormously costly and inefficient, and provides one of the man reasons for the development of EFTPOS (see Section 10.3.2.4).

The major problem for the banks is that an increase in the number of transactions does not necessarily increase revenue, but it directly increases costs. The reason for this is that revenue generated is related more to the number of account holders than to the volume of transactions. With increased competition from building societies (following deregulation), banks must look to reduce their operating costs, as well as attempting to increase revenue from other sources. The main target for these reductions has been, and will continue to be, in cutting branch operating costs. This could be achieved in a number of ways. First, there could be a reduction in the number of counter staff at each branch and the introduction of ATMs both within the bank and outside. The second step is a reduction in the number of branches in the network. In 1992, there was a network of 12 500 bank branches in the UK, a reduction of about 2500 since 1980 (Burton 1994). The third step, which is now being actively promoted, is the creation of hierarchical branch networks with fewer and bigger branches. Figure 10.3 illustrates the main features of this new structure. The regional head office has full responsibility for the banking services provided within its jurisdictional area. Area corporate offices located in city centres offer a full range of services to both private and business customers.

The operations centre provides the support services in the region, such as cheque-book and statement issue, processing of accounts, and so on. The service branches will provide primary banking functions (but not the full range of services provided by the corporate office) to small firms and private customers in urban areas. The rural branches, located say in market towns, will provide locality-specific services, such as loans for agricultural purposes. The effect of this restructuring is both a net reduction in branch numbers and an increase in the average distance customers will have to travel to consume particular services. There are therefore a number of important problems of spatial organization to address. First, to consider changing existing branch networks it is necessary to understand the current relationship between customers and branches. To what extent is a branch network necessary to retain customer loyalty as well as market penetration? Once this is understood, decisions on which branches to close and which to assign to the functions outlined in Figure 10.3 can begin to be made. As Section 10.5 will reiterate, model-based location analysis is absolutely essential in this process.

10.3.2.2 ATMs

As part of the push towards reducing costs of branch operations, cash-dispensing machines have been widely introduced by both the clearing

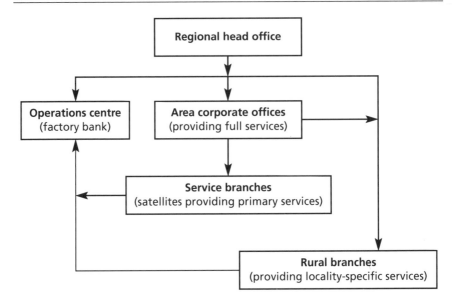

Figure 10.3 Evolving branch network system

banks and, more recently, building societies. Although cash-dispensing machines were introduced in 1969, ATMs did not appear until the mid-1970s. An ATM can be regarded as a cash dispenser with additional services, such as account balance, statement and cheque-book request, and so on. The trend will be for an increasing range of services to be made available from ATMs. The system of cash withdrawals is quite straightforward, and is depicted in Figure 10.4. Banks issue customers with cash cards and a personal identification number (PIN). The card can then be used in any appropriate ATM in any part of the country. The ATM is linked to a central computer that debits the account for the value of the transaction, and the cash is dispensed to the customer.

The ATM has the marked advantage for the customer providing cash-dispensing services outside banking hours, especially in the evenings and at weekends. For the bank it is estimated that ATMs can, if used as a substitute for cheque-based cash withdrawals, reduce transaction costs by up to 50 per cent. But this crucially depends on the substitution between cheque and ATM being achieved. This in turn depends, to an extent, on the location of ATMs. At up to £30 000 per installation, ATMs are capital intensive, and to repay this investment they need to generate substitution from counter transactions. This implies that they need to be in the right location for customers – which may not necessarily be where branches are located. If ATMs are used mainly in the evenings and at weekends, it suggests that they should be associated with residential or recreational locations more than workplace ones. To avoid the costly problem of massive ATM duplication, the banks have joined forces to

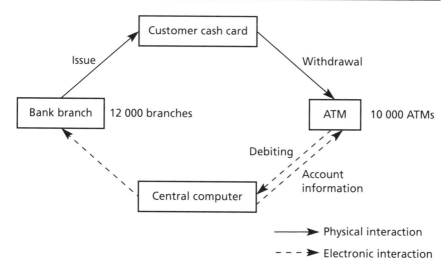

Figure 10.4 System of cash withdrawals from ATMs

provide shared networks which allow reciprocity of ATM usage. The two main networks are known as MINT (Midland, NatWest, and TSB) and Four Banks (Barclays, Lloyds, Bank of Scotland, and Royal Bank of Scotland). Additionally, many building societies are part of another shared network, known as LINK. However, the problem with reciprocity is that there is a cross-charging policy in place. If a Midland customer uses a NatWest ATM, the bank pays a small fee to NatWest. As a consequence banks have tried to prevent ATM 'promiscuity' by a proliferation of ATM locations.

Two major factors in site location will be the propensity for different types of person to use ATMs and the classic spatial relationship between supply and demand. If a very large number of ATMs are provided, widely dispersed, will this encourage the reluctant user to change from cheque to ATM cash withdrawal? Evidence suggests that there has been a significant change in attitude towards cash withdrawal. Consumers now view accessibility and minimum inconvenience as the major requirements. This contrasts with the situation 15 years ago, when one had to 'get to the bank on Friday'.

10.3.2.3 Credit and charge cards

The first credit card was issued by the General Petroleum Corporation of California (now Mobil Oil) in 1914. There are now over 1 billion credit cards world-wide, of which 60 per cent are in the USA. In the UK, Barclaycard (Visa) was introduced in 1966 and Access in 1972. In 1992, there were over 40 million credit cards in ownership in the UK, 26.5 million of which were operated by Visa or Access. Ownership varies substantially by social class, age, sex, and location.

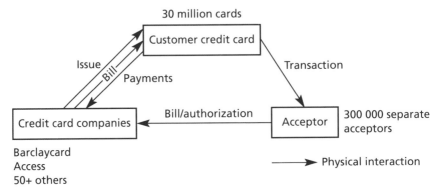

Figure 10.5 System of credit-card use

The system of credit-card use is illustrated in Figure 10.5. Credit cards are issued to individuals on application by the major credit-card companies from their central headquarters (for example, Access in Southend, Barclaycard in Northampton). Billing is centrally organized and issued by mail to card owners. The main categories of transactions are retail goods, petrol, and meals/hotels. The acceptor bills the credit card company for the value of the transaction, less commission which may range from 0.5 to 8 per cent. There may also be a telephone authorization for large transactions between the acceptor and the credit-card company.

As noted above, credit-card ownership has proliferated over the last 20 years. The rate of card issue grows at about 20–25 per cent per annum. Major recent growth has occurred in the in-store credit card, perhaps best illustrated by the phenomenal response to the Marks & Spencer card introduced in 1985. One major spin-off from in-store cards is that they allow a very detailed picture of customer profile and shopping habits to be built up by residential location, person type, frequency of shopping, and items purchased. This information itself will prove useful to retailers, and in particular will allow detailed store performance analysis and directed marketing to be undertaken, again using concepts from location analysis.

Specific spatial problems that are faced by credit-card companies relate to ownership and use. As we have seen, in some sectors of society ownership rates are comparatively low. This is also the case for particular areas in the country. In these cases, marketing efforts have to be targeted at those households with low ownership levels, and this will require the use of direct marketing methods, which will be discussed shortly. It may also involve increasing the number of outlets that accept credit cards for payment. This arises because of the aforementioned relationship between supply and demand. If non-owners perceive that credit cards are accepted in outlets they frequent, the possibility of their becoming owners is greater, everything else remaining equal. The decision by supermarkets to accept credit cards in their superstores is an illustration of this effect. Although the percentage commission the card companies negotiated with the main supermarkets was presumably very low, it was undoubtedly aimed at increasing card ownership (and use). In areas where ownership levels are relatively high, notably in the south east of the UK, the aim is clearly to increase use, both through provision of more participating outlets and through marketing and advertising aiming to convince the owner of the benefits of card use.

10.3.2.4 EFTPOS

As we have seen, one of the biggest drawbacks, from a bank's point of view, of cheques and credit cards is the cost of processing the transaction. This is largely due to the amount of paperwork involved. EFTPOS systems are aimed at avoiding the paperwork and directly debiting and crediting accounts. The process is illustrated in Figure 10.6. A bank issues a customer with a card and a PIN. The card can then be used at participating retailers for transactions. One major application area will be superstore food retailing, where bar-scanning systems read product and price. The customer pays the bill by inserting his or her EFTPOS card in a machine, entering the PIN on a keyboard. The funds are transferred directly by centrally located computers. The transaction time is estimated at 30 seconds, compared with approximately 2 minutes which it takes to transact by cheque. The cost of the transaction will be significantly cheaper than by cheque – approximately 10p compared with 25p.

Given that there were about 3.5 billion cheque transactions and 800 million credit-card transactions in the retail system in 1993 (Worthington 1995), the potential savings are enormous. Whether stores or banks are likely to benefit from this saving remains to be seen, although banks are likely to

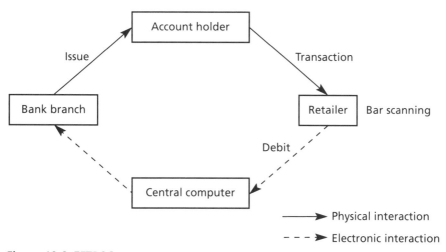

Figure 10.6 EFTPOS system

charge a commission for EFTPOS cards as they do with credit cards. The recently introduced CONNECT and SWITCH cards (which are not a true EFTPOS system) generated a good deal of friction between Barclays Bank and retailers. The major problem faced by banks and retailers concerning EFTPOS is persuading customers to use them instead of cheques. As cheques take several days to clear, the customer feels that she or he has a period of grace before his or her account is debited. Direct debiting may not appear so attractive. There is also a chicken-and-egg problem: if customers are to be persuaded to use EFTPOS cards, there must be a sufficient number of accepting outlets. The outlets themselves may well be reluctant to invest in the expensive capital equipment required for EFTPOS if customer usage is likely to be low.

10.4 FINDING THE CUSTOMER: SEGMENTING THE MARKET

10.4.1 Introduction

Some of the trends in product development, consumer awareness and consumption of personal financial services have been described. One major problem for financial institutions, given the nature of their products and the way in which they are consumed, is finding appropriate potential customers and persuading them to purchase. This is a common enough marketing problem, but is exacerbated in financial market places by the infrequency of financial service purchase and the products' relatively high value.

In previous eras, financial service selling was largely reactive, with customers providing the initial motivation for purchase. In the new era of competitiveness, selling is proactive in that the initial motivation is being generated by the supplier. As we have seen, this has important ramifications in the way that services are provided – the branchless bank could not have existed in the reactive era, but now this is a common form of service provision. This section looks at how methods of consumer differentiation have been developed to locate consumers, and in particular how geographical and demographic differentiation can be linked through the popular science of geodemographics.

10.4.2 Methods of differentiation and segmentation

There are six common methods that institutions, in general, use to attempt to differentiate between different types of consumer. These are demographic segmentation, geographic differentiation, geodemographics, product usage studies, psychographics, and needs/benefits analysis. This section focuses on the first three, which can most conveniently be used for analysis within the financial services sector, and briefly mentions the others (see also McGoldrick 1990, Birkin 1995).

Table 10.2 Ownership of extra interest accounts by life cycle (*Source*: adapted from NOP Financial Research Survey 1987)

Life cycle	% of all accounts	% of sample
16–24, single	5.8	15.2
16–34, married, no children	4.2	4.6
25–54, married, with children	18.9	22.6
45–64, married, no children	32.9	20.3
65+	21.8	18.5

10.4.2.1 Demographics segmentation

The study of the personal and household statistics of customers identifies specific lifestyle types based upon such factors as age, socio-economic group, occupation, and household composition. Life-cycle stage can be seen to influence the take-up of financial services more than any other single factor. For example, Table 10.2 presents the ownership of extra interest accounts by lifestyle stage. These can be strongly biased towards older households without children, whereas bank loans (Table 10.3) demonstrate quite a different pattern, predominating in younger households with children.

Demographic segmentation systems have a number of advantages in the identification of potential customers. Most notably these are as follows:

- They are relatively easy to implement within most financial institutions, as these already collect the basic demographic data, including age, sex, occupation, and household composition.
- These systems are easy for operating staff to understand, as basic demographic analyses are commonly used in marketing at present.
- Demographic analyses and, in particular, the 'life-cycle' type of analysis, noting age and household status, have an immediate relevance to financial 'need' and, hence potential sales strategy.
- Demographics is commonly used in the presentation of secondary data, both research and government data sets, and hence any targeting developed along demographic lines would be cross-referenced.

Table 10.3 Ownership of bank loans by life cycle (*Source*: adapted from NOP Financial Research Survey 1987)

Life cycle	% of all loans	% of sample
16–24, single	14	15.2
16–34, married, no children	11	4.6
25–54, married, with children	38	22.6
45–64, married, no children	17	20.3
65+	1	18.5

- Most media owners identify their products' appeal in terms of demographics and, hence, a targeting system based on these criteria would assist in advertising/media strategy.
- Demographics identifies the 'type' of person in the target market segment and, thus, could assist with creative planning and strategy.

Demographic segmentation systems have several disadvantages for financial segmentation, however, notably as follows:

- There is no insight provided into the motivation to purchase amongst the target groups. This could result in major difficulties in the creation of the creative approach to the target groups.
- The demographic segmentation system cannot be replicated to establish non-customer prospects, as there is no link between demographic 'qualities' and the Electoral Register – the biggest source of 'cold' prospects. There are a limited number of lists which are segmented by demographic attributes in order to provide non-customer prospects.
- In general, demography cannot be used to analyse particular catchment area potentials or assess spatial concerns.

10.4.2.2 Geographic differentiation

The geography of product purchasers simply identifies customer types by their location (either residential or workplace) or by the location of the purchase point. Thus, a simple division of north/south can be established, or more complex county, ward, postal district, or post-code areas defined, dependent upon the segmentation requirements. Geography has traditionally been an important differentiator in financial markets due to the history of product development. Life assurance, for example, has strong connections with Scotland and the north of England, while credit-card usage remains dominated by the south (Table 10.4).

Geographic segmentation systems have key advantages for financial marketing. Notably they are:

- easy for most financial institutions to implement and for the various companies' operational staff to understand/use, as a customer's residential address and branch address (where applicable) are a feature of most financial companies' current databases;
- totally compatible with most companies' management information systems (MIS);
- readily comparable with secondary research and published data, such as the Financial Research Services (FRS) survey, and government statistics (CSO);
- geographically referenceable – that is to say, target segments could be matched with the Electoral Roll in order to establish 'cold' target market

prospects, geographical areas could be considered in terms of existence of the target market, and so on.

However, the systems have several drawbacks for financial institutions, notable as follows:

- A geographical analysis cannot identify the 'type' of customer buying the product. The population in the area defined is considered to be heterogeneous and hence its specific, individual banking needs cannot be approached directly.
- Other than in the broadest sense, a simple geographical segmentation cannot assist in the definition of advertising or promotional strategy.
- There is no insight provided into the motivation to purchase amongst the target groups. This can result in major difficulties in the preparation of creative or promotional material, which should be designed to appeal specifically to the target audience.
- Most information collected from market research survey allows only a coarse spatial label to be attached, such as 'county', 'TV region', and so on. For many purposes, much finer spatial detail is required.

Clearly some of the deficiencies of geographic differentiation are the advantages of demographic segmentation and vice versa. An obvious next step is to link these together, through geodemographics.

10.4.2.3 Geodemographics

As seen in Chapter 4, geodemographics is the integration of demographic and geographic information to provide detailed profiles of the characteristics of areas, ranging from EDs to larger units. Most commercially available geodemographic systems (such as ACORN, PiN, SUPERPROFILE, and

Table 10.4 Geography of product purchasers (*Source*: NOP 1987)

Standard region	% using a credit card in last month	% holding life insurance
Scotland	52	46
North	37	45
North west	54	38
Yorkshire and Humberside	56	36
West Midlands	54	31
East Midlands	49	33
East Anglia	43	28
Wales	48	30
Outer London	59	32
Inner London	57	35
South west	54	34

MOSAIC) are generated by the use of clustering techniques, such as principal components analysis, applied to Census data. These methods attempt to isolate components that are identifiable as distinct groups and then to assign EDs to one of these categories. It is also possible to integrate market research survey information to relate demographic groups to product types. For financial institutions, geodemographic targeting systems such as these have many advantages, such as the following:

- Implementation is easy. The geodemographics 'code' is traditionally allocated to individuals by reference to their residential post code.
- Because of the geodemographic link with addresses and post codes, a geodemographic segmentation system would be totally compatible with most financial institutions' MIS.
- Geographical referencing enables customer prospecting, cross sales, and customer recruitment to be undertaken.
- Geodemographic segments can identify specifically the 'types' of person purchasing the products. This can assist with the preparation of creative plans and strategy.
- Many large research surveys, such as the FRS surveys, have their results cross-tabulated by geodemographic criteria, hence providing the facility to cross-analyse in-house performance with research findings.
- Many media owners identify their product appeal in terms of geodemographic criteria and, hence, geodemographic systems can be used to inform advertising/communication decisions.

Geodemographic systems do, however, have some disadvantages. These include the following:

- Some systems have many segments, which may be difficult to understand/relate to.
- Whilst geodemographic systems can be combined and used in different ways, it may be difficult for the various operating staff, unused to geodemographic systems, to achieve flexibility through segment manipulation.
- In general, geodemographic segmentation systems do not incorporate any aspects of customer motivation to purchase. As discussed earlier, this can cause creative problems.
- A major problem exists in relation to the categorization of areas. Each area (ED, postal sector, etc.) is assigned to a particular cluster group using the average customer profile characteristics. However, there may be significant numbers of households and individuals that do not fall into this average categorization – for example, ACORN group Y37, classified as 'private houses – welloff-elderly', contains 65 per cent of individuals who are less than 65 years of age. Targeting using these methods is therefore likely to miss large sections of the market, unless detailed training can minimize the abuse of geodemographic indicators.

It is in relation to this last point that we advocate the use of microsimulation methods for population reconstruction. This methodology, fully described in Birkin and Clarke (1988) and Clarke and Holm (1987), allows for a reconstruction of the population (households and associated individuals) for small areal units such as EDs at the micro-level, and for the incorporation of an extended list of attributes, including those derived from market research surveys (see again Chapter 3). For example, it is possible to generate attributes such as income, expenditure, and financial service usage, as well as conventional demographic variables. This then allows for the enumeration of various types of household in specific areas – say, the number of credit-card owners with income over £20 000 per year who live in two-car-owning households. The methodology also has the important feature of being updatable, thus releasing institutions from making investment and marketing decisions based on information six years old. The method can also be applied to customer databases to provide best estimates of the categories of variables that are not known but would prove useful in targeting.

10.4.2.4 Other methods

Three other methods for segmentation and differentiation are worthy of mention, though none so far has been developed in a spatial context.

Product usage studies consider which of the company's products any individual customer or household has. This can often be readily ascertained from customer records files. Linked with geodemographic referencing, it could prove a useful additional segmentation tool.

Psychographics considers consumer attitudes and motivations for purchasing and owning products. Like commercially available geodemographic systems, psychographic systems such as Stanford Research Institute's VALS (Values and Lifestyles) attempt to categorize households into particular groups on the basis of a form of cluster analysis. The potential obviously exists for adding a spatial dimension following the procedure developed for geodemographics (geopsychographics?), but perhaps more interestingly the results could be used as inputs into the microsimulation models discussed above.

The final segmentation system is known as needs/benefits analysis. This involves identifying why customers are buying various products and services, which normally involves extensive market research and is thus costly to implement. It does provide institutions with valuable information about motivation, but can suffer from subjective interpretation. For more detail on these methods see Birkin (1995).

10.4.3 Summary

The use of segmentation systems in the analysis of financial services and in particular the identification and differentiation of consumers is an important process. It has two major roles to play in terms of spatial analysis. The first is in the field of direct marketing, where a target market has to be identified. The second is in the analysis of site location. Section 10.5 demonstrates how geodemographic systems can be combined with modelling methods in a number of application areas.

10.5 GIS, GEODEMOGRAPHICS, AND SPATIAL ANALYSIS

10.5.1 Introduction

It is clear from the above discussion that geodemographics have played a major role in the development of market analysis in the financial service industry. It is perhaps not surprising to witness the rise in the use of GIS to support such analysis. Indeed, a large number of geodemographic vendors have integrated their systems with GIS functionality, especially visualization. Grimshaw (1992) found that 34 per cent of all building societies, and 52 per cent of the top 30 societies have purchased or are purchasing a GIS. Their use seems to be limited to database management and mapping, and almost all were linked to a geodemographic package, with ACORN and Pinpoint being the most popular. Thus, 80 per cent of users were employing GIS for target marketing purposes.

This section examines three areas where the application of GIS, geodemographics, and methods of spatial analysis can contribute to the marketing and planning of financial services provision. The areas selected for illustration are:

● branch evaluation and rationalization;
● television advertising optimization;
● direct mail targeting.

They are discussed in turn.

10.5.2 Branch evaluation and rationalization

Traditional retail theory views the localized retail branch networks of banks and building societies as a necessity for customer service: here customers can view and access all the available products and services, including the advice and expertise of the staff. Developments in sales and distribution techniques within the financial services market place are changing this view. Retail branch networks are a major cost centre for all financial institutions. In the interests of cost efficiency, branch networks are being increasingly closely monitored in terms of market share, market potential, and performance in relation to financial standards and to other branches in the network.

The opportunities for individual financial product sales and the requirement for customer/product servicing vary not only by product but also by neighbourhood, because of differences in local-area characteristics. It is therefore important for companies to establish a consistent standard upon which to evaluate branch performance. Geodemographic analyses can assist in this task, by considering the profile of potential customers in a catchment area and relating it to the profile of the actual branch customers.

In addition, the presence or absence of competition in an area will significantly affect both branch performance and potential. Analysis of one institution's branch network cannot be successfully undertaken in isolation from those of others. However, although institutions will have very detailed information on their own customers, they will have little or no information on those of their competitors. It is nevertheless possible to use models to infer behaviour of non-customers from the limited information obtainable from a number of sources. For example, it may be reasonable to assume they behave in similar ways to a bank's own customers with the same characteristics.

It would also appear that GIS could assist with traditional branch location analysis, employing the methods developed for more general retail analysis (cf. Chapter 4). Indeed, Howe (1991) offers a classic example of buffer and overlay procedures for estimating the revenue of a potential new building society branch in Truro, UK. We have already discussed the problems with this sort of analysis in Chapters 3 and 4. There is an additional problem in the financial service sector. One important feature of financial services branch networks is the ability of one branch to support the activity of another. This occurs because, when a customer opens an account in a particular branch, he or she can transact on that account at any of that bank's other branches or ATMs. Because convenience is now such an important factor in transacting, the availability of a branch near a customer's residential location may increase the probability of that customer opening an account with a city-centre branch of the same bank because their overall access to transacting facilities is high.

We term this phenomenon the 'network effect' – market share is enhanced in a non-linear way through the development of a branch and ATM network in a region. This is illustrated schematically in Figure 10.7, where the relationship between market share and number of branches is plotted. If a bank has a small number of branches in a region, market share will remain low because of the low density of access points.

However, beyond a certain level the network effect takes off and market share grows rapidly, up to a point where the branch network is saturated and new branches simply steal customers from existing branches in the same network. Clearly, it is important for any financial services organization to know where in this curve it lies in each of the regions in which it has representation.

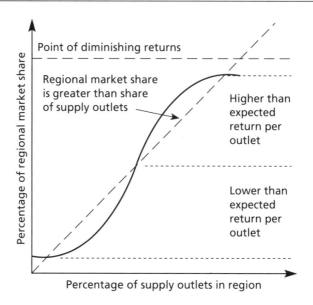

Figure 10.7 Network effect of bank branches

Figure 10.8 (Plate 10.1) provides an example of this network effect at work. This map of Greater London postal sectors shows the residential location of account holders at the Moorgate (City of London) branch of a large bank. Also overlaid on this map are the locations of the bank's other branches in the London region. It is clear that the peaks in their accounts by postal sector often coincide with locations of suburban branches.

Given the existence of these factors, it still proves possible to generate a set of models and performance indicators for branch evaluation, but it is necessary to differentiate between account holdings and business transacted. The procedure requires the articulation of the levels of service demand in residential districts and the identification of all outlets providing that service in an area. The former can be achieved through the use of geodemographics, either incorporating information from a profiling system such as ACORN or MOSAIC or using the Census data combined with market research data described earlier (see Chapter 5). It is then possible to calibrate a spatial interaction model, using customer data on residential and branch location, for both account holdings and transactions. This will allow measures of potential performance to be compared with actual performance, by service type for each branch in the network, as well as the calculation of other indicators such as market share and market penetration. The calibrated models can then be used for the assessment of new branch location or rationalization of existing branches. An example of such modelling is provided in Section 10.6.

One area where this type of analysis will become more important is in assessing insurance broker location. The 1987 Financial Services Act forces brokers either to sell all products on a value-for-money basis or to act as a tied broker selling, exclusively, one company's products. Because of the legislative implications of the former, it is anticipated that most brokers will take the exclusivity route. As a consequence, it will be vital for the insurance companies to have a network of tied brokers that allows them to achieve the required levels of market penetration. Non-representation in areas will in future result in very low levels of penetration compared to the current situation.

10.5.3 Television advertising: optimization of resource allocations

With the growing competition in the financial services market place, financial institutions are relying increasingly upon the traditional advertising techniques of the retailers to sell their products. Advertising by banks, building societies, and other financial institutions has grown dramatically over recent years, as shown in Table 10.5. Much of the increase in volume of advertising has been accounted for by the proliferation of television usage by the financial institutions concerned. At up to £100 000 per peak-time 30-second slot, the cost of advertising campaigns can rapidly rise; hence the need for considered analytical approaches to campaign planning both by television region and by particular airtime 'slot'.

Decisions regarding product advertising in terms of both television area and individual slot choice are complex, due in the main to the proliferation of pricing arrangements for television airtime in the UK. However, at its base is a simple concept that the value of airtime to an advertiser is dependent upon the number of 'target' market individuals who see the advertisement; hence the advertising measure of OTS or 'opportunities to see'. OTSs are highly dependent upon such factors as area size, area characteristics, and advertisement slot positioning (which programmes come before and after the slot).

Area size and characteristics and the consequent 'potential' demand for financial products and services in any area can be identified for each of the 13 television regions by the use of geodemographic analyses and research data or the microsimulation methodology described earlier in this chapter. Hence the advertiser would, by the use of these techniques, be able to calibrate an area 'potential' index and inform network allocation models. Broadly based resource apportionment models of this kind are being used quite extensively by financial advertisers in the UK today.

Geodemographic analysis of the viewers of various programmes, rather than the simplistic age and socio-economic group data currently produced by

Table 10.5 Growth in advertising expenditures (*Source*: NOP 1987)

Financial advertisers	1980	1983	1986
Building societies	20 949	62 815	68 593
Credit cards	4 548	11 156	16 742
City and financial services	8 068	21 046	42 287
Foreign/overseas banks	5 195	6 630	7 888
Insurance/assurance companies	12 494	35 223	62 896
Joint stock banks	17 859	44 851	68 135
Unit trusts	2 898	8 928	25 624

JICNAR, could assist advertisers at a much more finely tuned level to define particular airtime slots which deliver the maximum target market audiences. Total resource allocation optimization models could thus be defined.

10.5.4 Direct mail: targeting using GIS and geodemographics

Direct mail is the dispatch of unsolicited letters to either customers or non-customers with the object of selling them a product or service. The crucial factor in cost-effective direct mailing is to direct the letters toward those households most likely to buy the product/service. Direct sales costs are virtually fixed whatever the 'take-up' rate, and hence any improvement which can be made in the targeting of mail items is to the direct advantage of the mailer. Geodemographic analyses can play a major part in the cost-efficient targeting of mail items.

Direct mail is becoming an increasingly popular method of selling products, and financial products are no exception. In 1985–86, it was estimated that 13 per cent of all direct mail was financial in origin, and this percentage increased during the late 1980s and early 1990s. In 1986, financial direct mail was estimated to be worth around £60 million. Thus optimal targeting systems could have a major impact upon the cost-efficiency of financial direct marketing.

Direct mail can be sourced in four ways:

- existing customer listings;
- affinity listings (purchased customer lists of a similar 'type');
- lifestyle databases;
- Electoral Register.

In each of the four, geodemographic analysis can play a major role in the cost-efficient targeting of mailings. Existing customer lists are the best source of names and addresses, since a bond of trust already exists between the potential purchaser and the seller. However, not all existing customers may be suitable candidates for the products being sold by the company – an individual heavily over-extended with credit would not, for example, be a good customer to mail with further personal loan items. The art of mailing profitably to an existing customer base is to know your customer, and as a consequence many financial institutions are creating complex, computerized customer records, from which they can select and extract the contact names most suitable to new product sales. Geodemographic indicators frequently play a major part in these customer databases, permitting a simple extraction procedure incorporating both customer demographics and geography to take place. A good example of such a product is Pinpoint's FINPiN system. This has been incorporated into a GIS so that the client can enjoy all the added benefits associated with graphic displays and windowing/zooming, etc. FINPiN identifies a number of client groups based on local geodemographic profiles. These groups include:

- financially active (will have a current account, savings, mortgages, and investments);
- financially informed (will have perhaps two or three of the above: company-loyal);
- financially conscious (perhaps with just a current account and mortgage);
- financially passive (perhaps just a current account: possibly none).

The establishment of the geodemographic targets on the base of which extractions can be made could be undertaken by recourse to an analysis of existing customers or by consideration of research data. Affinity listings – purchased name and address listings from other companies, such as Amex, book clubs, etc. – are increasingly being segmented by geodemographic classification systems in order to permit further precision in targeting. All affinity listings will require further analysis before mailing in order to eliminate the existing customers of the company from the mailing list.

Lifestyle databases have recently provided a new way of identifying target households for financial services' product offerings. In the UK, there are three main lifestyle data companies: NDL, CMT, and ICD. They all collect their data in slightly different ways, but all of it is voluntarily provided. Individual householders fill in questionnaires concerning social and demographic attributes (age, sex, occupation, tenure, etc.) along with lifestyle data such as sporting and leisure interests. These databases have been assembled steadily over several years and the largest (NDL) contains almost a 50 per-cent sample of UK householders. The beauty of this information is that it relates directly to individual households, not to the characteristics of residential areas such as postal sectors or Census wards.

The Electoral Register is the most comprehensive listing of individuals' names and addresses in Great Britain today. Mailing lists can be obtained directly from the Electoral Register by various different criteria, including household status, age of household head, age of children, etc. But geodemographic analyses remain the most popular. Several institutions provide electoral selections by geodemographic type by matching Census ED geodemographic characteristics to postal areas and electoral listings via a grid reference matching and centroid proximity exercise for customer use. Again, all electoral listings require further analysis to eliminate existing customers from the mailing list.

10.6 MODELLING THE FINANCIAL SERVICE MARKET

10.6.1 The modelling process

To illustrate how the methods of market analysis described earlier in this chapter can be applied in a financial services context, we report on a project undertaken for a GMAP client in the city of Adelaide, Australia. The model that has been built has a number of objectives:

- to predict the number of new accounts being opened over a 12-month period at each financial centre in Adelaide and at each client and competitor branch;
- to predict the number of new accounts being opened over the same period by residents of each postal sector for the client and its competitors;
- to predict where new accounts opened in a financial centre come from.

A variant of a spatial interaction model was used to achieve this. The modelling process proceeded in four stages, as follows:

1 *Estimate the demand for new personal account products in each postal sector.* This was achieved by combining Census data with market research data supplied by the bank. The market research data provided the probability of different types of person opening new personal accounts by age and social class. The Census provided the number of people in these categories for each postal sector. Suppose the likelihood of a 16–24 year-old in social class 1 opening a new account in a 12-month period is 17 per cent. Then the number of new accounts expected from that age and social class is 0.17 multiplied by the number of that person type in a postal sector as in the Census. The total for the postal sector would then be the sum of all population groups multiplied by their respective probabilities.
2 *Identify financial services centres.* Banks and building society branches tend to be clustered in centres which also contain a range of other retail outlets. Using the unit post code of each branch in Adelaide, it is possible to identify these clusters through visual inspection.
3 *Allocate residential postal-sector demand to centres.* To allocate demand to supply, we built a spatial interaction model of account-opening

behaviour. The modal allocation works on the following principles (which should by now be familiar to readers from Chapters 4 and 5 and subsequent illustrations):

– *centre attractiveness*, determined by the size and quality of each financial service centre. This is calculated by identifying the number of branches in a centre and their size, together with a measure of the brand strength of each bank present in the centre;

– *centre accessibility*, a measure based on drive times between postal sectors and centres.

In simple terms, the greater the relative attractiveness and/or accessibility of a centre the greater the flows between a postal sector and that centre. To determine the relative strengths of these two factors, the model has to be calibrated in the usual way, in this case using client customer data. The base model reproduces existing customer data as far as possible. Once all the flows from each postal sector to each centre have been allocated, it is quite straightforward to calculate the total estimated expenditure in any centre by summing over all origins.

4 *Allocate total business in each centre among competing outlets.*

Stage 3 generated an estimate for each financial services centre of the total number of new accounts being opened in that centre. We then needed to convert this into a prediction for accounts opened at each branch in the centre. The amount of information available for each outlet in a centre is variable. For the clients' branches, we have data on the size of each branch, the number of staff, and so on. For competitors, we know very little other than the brand name of the company operating the branch.

Each competitor was assigned a 'competitiveness' weighting based on its share of the new account market in Adelaide in relation to its share of all bank branches in the city. This gives a general measure of its ability as a brand to capture more or less than the expected share of the market. Using this brand strength measure, we assigned the total market in a centre amongst the branches present on the basis of their brand strength. Although a relatively crude approach, Stages 3 and 4 combined produced remarkably good results for the clients' branches. Over 80 per cent of the predicted revenues for branches were within plus or minus 10 per cent of the observed.

Once the model had been developed, it could be used in a variety of ways. Figure 10.9 (Plate 10.2) shows how it is possible to estimate market penetration for the client. After studying present performance, the bank can then undertake a series of 'what-if' scenarios. We can illustrate this through a single case study from the Adelaide system.

By examining Figure 10.9 (Plate 10.2) and relating this to where the clients' branches are located, it is possible to identify obvious gaps in their network – where market penetration is low and there is an absence of branches. Figure 10.10 (Plate 10.3) displays this information for a region around the suburb of St Agnes. As can be seen, the bank's market penetration is low in several postal sectors, and there is no branch in St Agnes. Using the model, we can assess the impacts of opening a branch in St Agnes. Table 10.6 presents the results of running the model. We can see that 680 new accounts would be generated by the branch, but of this number 76 would be cannibalized from the bank's other branches in nearby centres. The net effect of 604 new accounts could be taken as a starting point from which to examine the branches' viability on financial grounds. As can be seen from Figure 10.11 (Plate 10.4), the impact on market penetration is quite dramatic, with the obvious gap in market share eliminated. Tactically, this rectification of a clear geographical weakness might be as important to the bank as the financial equation.

10.6.2 Towards Integrated spatial analysis

This chapter has described three areas in which spatial analysis and geo-demographics can be used to assist in the marketing and analysis of financial service provision. Currently, in many situations these three areas would be seen as relatively distinct operational functions. There does seem obvious merit in drawing together these three areas (and possibly others) to produce an integrated marketing and analysis function. For example, if television advertising is advocating targeted consumers to purchase a product or service, it would make commercial sense if consumers had readily

Table 10.6 Impact of new transaction account business of new branch in St Agnes, Adelaide (*Source*: GMAP 1994)

Branch	Impact
Modbury Central	−13
Tea Tree Gully	−10
Clovercrest Plaza	−7
Farview Park	−6
Golden Grove	−5
Salisbury	−5
Ingle Farm	−4
Pava Hills	−3
Nailsworth	−2
All other branches	−21
Total lost to other branches	**76**

Note: The new branch is predicted to sell 680 new transaction accounts per year. However, the net gain to clients is 604 new accounts.

accessible branches where the product could be sampled and discussed. If consumers cannot translate the interest generated from the advertisement into a visit to a branch supplying the service, the advertisement is wasted, and response and conversion rates will remain low. Yet we regularly witness television advertisements from financial institutions that either have a low density of branches in a region or have a concentration of branches in one area of a region.

Integrated analysis would link media advertising and direct marketing with branch location analysis. This could result in the replacement of television advertisements with much more spatially targeted media campaigns, such as local press and radio and/or the extension of the branch network into areas not currently served. In addition, the use of predictive modelling could aid in the evaluation of the effectiveness of the advertising/marketing campaign.

For example, if it was estimated that the campaign should improve market penetration by 5 per cent, a spatial interaction model could be used to predict where that increased market penetration should manifest itself branch by branch. By monitoring branch sales, the actual impact of the campaign could be measured against its intended effect.

10.7 CONCLUDING COMMENTS

We have aimed to illustrate the nature of financial service organization and provision in the UK and the way in which spatial organization and planning are key features of these systems. By picking a number of examples, we have shown how model-based analysis coupled with geodemographic techniques can assist financial service institutions in improving their efficiency and effectiveness.

It should also be mentioned that, on discussing the contribution of analytic methods in the financial services arena, we have deliberately avoided many other issues that will be of interest to geographers, such as the trends in employment, particularly female and part-time labour, the influence that financial institutions have on the functioning of the economy, and the growth of personal credit. Despite the sterling attempts of some (such as Daniels 1985), there remains further potential for geographers to contribute to financial services research on a number of different levels.

CHAPTER 11

Conclusions

11.1 INTRODUCTION

The preceding chapters have argued for a more integrated approach between GIS and model-based analysis for market planners in a wide range of service and retail activities. We have provided a corresponding range of examples to illustrate the importance of this argument in many areas of business activity. The argument has come from our collective experience in working with business executives. They seem increasingly less prepared to accept the vendor-driven functionality within GIS which seems to force clients to shoe-horn their geographical problems into the black-box functionality of proprietary GIS. We have argued that GIS analysts must understand modern business problems far more if they are going to offer business solutions. We suspect that most large organizations still undertake their geographical analysis and planning without any formal use of intelligent GIS tools. In fact, most chief executives and company directors are still unaware of what this capacity can provide. There is therefore a need to educate potential users of such technology about the benefits it can provide. Similarly, GIS proponents need to be educated in the needs of business. Spatial analysis will only be used in business if there are good reasons for so doing. We hope we have demonstrated that such problems are complex and demand analytic functionality which can cope with that complexity.

11.2 QUANTIFYING THE BENEFITS OF IGIS

In the GIS literature to date, there have been few attempts to undertake a traditional cost-benefit assessment of the use of GIS (though see Dickinson and Calkins 1988, Smith and Tomlinson 1992). Where there has been such an attempt, these benefits are usually related to improvements in the efficiency of data storage and handling and the development of corporate sharing of information and resources. It is very difficult to quantify the financial savings associated with such benefits. Yet, for business users in particular, it is essential to be able to estimate some kind of substantial financial benefit to be derived from spatial analysis if we are to convince them to make significant investments, since the production of maps *per se* is not an activity they would normally engage in. Wellar (1993: 17) offers the following logic test: 'If, after study, the evidence suggests that a GIS means Better Data – Better Information – Better Decision Processes – Better Decision Outcomes – Better Bottom Lines, then, buying or expanding a GIS makes business sense and deserves favourable consideration.'

In the following sections we give three case-study examples of the sorts of benefit users can enjoy if they adopt GIS and spatial modelling methods, or IGIS as described in the chapters above. It is important to emphasize to business users through such examples the impact on the bottom line.

11.2.1 The auto industry

The auto industry is a sector that has long recognized the importance of geographical planning and analysis. All the main auto manufacturers distribute their products to the market via a network of franchise dealers. As we saw in Chapter 9, these dealers are independent businesses but are allocated an exclusive geographical territory, to which the manufacturer agrees not to assign any other dealer subject to the existing dealer meeting certain performance criteria. Clearly, most manufacturers aim to maximize their market share and profitability in the market. From analysis of the voluminous amounts of registration data, it is clear that there is a very strong relationship between market share and dealer location. In other words, the more dealers the manufacturer appoints, the greater the likely market share. However, this is traded off against the fact that, as market share increases, there are diminishing returns and the sales of each dealer reduce, thus affecting individual dealer profitability and possibly the scope for retail price discounting. As a consequence, manufacturers are trying to find a balance between maximizing market share and at the same time ensuring that each individual dealership is a profitable business in its own right. Achieving this balance requires a thorough understanding of existing market performance, and the ability to examine alternative scenarios through an IGIS approach.

The particular example that we will use to illustrate the benefits of this type of approach is based on the analysis of the Toyota network in Seattle, Washington State, USA, which we discussed in Chapter 9. The specific question that we addressed was whether there are possibilities of extending the Toyota network in Seattle in such a way as to generate incremental sales and market share, without detrimentally affecting the viability of existing Toyota dealers, and at the same time creating successful new dealerships in their own right. We argued that these sorts of question demand a type of analytical solution, known as an optimization model. Although these models are well known to quantitative geographers, they are generally not available in proprietary GIS. Using these we identified two different optimization solutions. One solution guaranteed that any new location should have a miniumum of 450 cars per year and a location more than 13 minutes' drive time away from an existing Toyota dealer. In this case the solution identified two dealers, which would generate incremental sales of almost 750 Toyotas in the Seattle region. The second solution required any new location to have a minimum of 600 sales but a location at least 10 minutes' drive time from an existing dealer. In this case a solution was identified, but the locations were markedly different from the first example. However, almost 1000 incremental sales were generated in this second example. Based on 1000 incremental sales and given an estimated profitability of $1500 per vehicle, the potential benefit to Toyota is in the region of $1.5m per year in this single geographical market. Given that there are in the order of 80 markets of a similar size to Seattle in the USA, the exercise repeated across the country could generate benefits of well over $100m. Even if the figure was only 10 per cent of this, the benefits would far outweigh the costs of developing such a system.

11.2.2 A major DIY retailer

The second example that illustrates the benefits of using IGIS relates to a major DIY retailer in the UK. Although we have not used this example directly in the retailing chapters, the analysis undertaken using IGIS is typical of that provided to all the retail clients.

This particular DIY retailer has a substantial network of outlets throughout the country and has a stated aim of increasing its market share in the DIY market. However, it faced a major problem in being able to predict accurately the revenues that would be generated by opening a store in a specific location. The historic accuracy of its revenue predictions were in the range of plus or minus 30 per cent of actual revenue generated at a site when opened. This wide variation in forecasting accuracy essentially meant that the retailer had to predict a revenue of at least 30 per cent above the break-even point to be assured that the new store at that site would be viable. As most people involved in retailing would appreciate, there are not many opportunities to identify sites where revenues are so great as to return a 30 per-cent-plus level of profitability. The consequence of this for the organization has been that they have not been able to develop their store network at anything like the pace they would wish.

The importance of spatial models is their ability to predict new store revenues closer to a 10 per-cent accuracy level in most business sectors. With such greater accuracy, revenues for the new stores can be calculated within narrower confidence limits. The benefits that this system generates for the retailer are quite easy to calculate. As well as having timely access to relevant market information and the ability to manipulate this to understand the markets and the retailer's performance in it better, the major benefit is related to investment decisions. Let us suppose the organization were to open five new

Table 11.1 Financial return over five years

Year	1	2	3	4	5
Benefits *Total £37.5*	2.5	5.0	7.5	10.0	12.5

Note: All figures are in £millions.

stores per year over the next five years. These stores would not satisfy the plus-30 per-cent profitability that would have been needed with the previous system, and could therefore be seen as stores which the organization would not have otherwise developed. Let us further assume each new store generates a revenue of £5m per annum and has an average net profitability of 10 per cent. Table 11.1 illustrates the flows of profitability over this period of time. The total sum of £37.5m is over 100 times the total cost of the SDSS or IGIS developed by GMAP for this client. This gives a fairly clear indication of the substantial business benefits the client is receiving from investing in IGIS.

11.2.3 A major UK building society

Finally, we use an illustration from the financial services market. In this case the benefits do not come from new store openings. This particular organization has a network of several hundred branches throughout the UK, although there are significant regional concentrations. The issue that the organization was concerned with was the performance both of the brand and the individual branches within the M25 area of south-east England, mainly in the London conurbation. The organization was considering investing several tens of millions of pounds in refurbishing its hundred or so branches in this region. It was believed that the poor performance of the branches and the brand could partly be explained by the relatively poor physical fabric of the branches.

However, before embarking on this programme, it was deemed appropriate to delve deeper into the real causes of the problem. This involved the construction of a national information system that contained detailed information on the demand for several different financial services products by small area across the country, estimated by marrying information on population types (age, sex, social class, etc.) with their propensity to hold accounts of certain types (high-income deposits, mortgages, etc.). We also identified the location of every single bank and building society outlet and assigned these branches to financial services centres, of which we identified over 2000 in the whole of the UK. Using the organization's customer data, it was possible to calculate the market share that was achieved in each centre, and the market penetration it had achieved in each small area for the whole product range.

Once the IGIS had been designed and built, it was used to perform a 'critical success factor' analysis, which attempted to tease out, from the information and the model output, what factors caused good performance in certain regions and what factors generated poor performance. It turned out that three main factors appeared to be at work. The organization tended to perform poorly when the following conditions occurred:

1 a low level of population per branch in a region;
2 a high level of representation of its main competitors in the region;
3 high house prices and average mortgage values in a region.

Our analysis demonstrated that the significant investment that was to be made in refurbishing the branch network in south-east England would not address the main structural problems being experienced there. Furthermore, we identified opportunities in other regions of the country where the critical success factors were in evidence but the organization had an under-developed branch network. Our recommendations were that the organization should not go ahead with its multi-million-pound investment in refurbishment, but rather that a programme of network development in suitable regions be pursued, coupled with an examination of the possible acquisition of a smaller financial service organization in a suitable region.

In this example, the quantifiable benefits are not so much the profit streams to be generated by new developments through the application of the IGIS as the significant savings that have been identified through not proceeding with their planned investment. The application also demonstrates that development of an IGIS approach can provide key inputs at a senior level within an organization, and not just assist with tactical operational issues.

11.3 CONCLUDING COMMENTS

It can be argued that the analysis of geographic information helps organizations by adding value to their data and information. Figure 11.1 attempts to represent a value chain that is required by most organizations. At the foot of this diagram are data, which probably have little value in their own right unless they can be turned into information. In the past, it was often argued that most companies were data-rich but information-poor. The development of management information systems, GIS, and the like have now resulted in a situation where many organizations are information-rich but intelligence-poor. We have argued in this book that it is through the embedding of geographical modelling methods that information can be transformed into intelligence. However, it is also true from many of our experiences that, even when we get as high in the value chain as

Figure 11.1 Value chain

intelligence, there is no guarantee that organizations will use this intelligence to plan their businesses better. There is, therefore, an important role for geographers, namely to operate in what might be thought of as the traditional management consultancy role of converting intelligence into proposed action plans. This is a territory that most players in the GIS business are unfamiliar with, and that in itself presents new and interesting challenges.

From our own experience with GMAP, we have over the last two years faced an increasing number of requests for the provision of strategic information and advice from very senior levels within organizations, but without the full GIS package being delivered to and supported by that organization. We detect that there is a possibility of a significant divergence in the fall-out from the GIS revolution. On the one hand, there is the development of the GIS commodity platform, which sits alongside other desktop computing tools such as spreadsheets and databases. Here GIS becomes simply another operational tool for middle and junior managers. In this environment, mapping supports day-to-day activities at the lower end of the value chain depicted in Figure 11.1. On the other hand, there are the more sophisticated intelligent GIS or IGIS outlined in this book, which help to put together a strategic advice package which provides inputs at the highest level within organizations. In the former case, GIS would be developed as a commodity, be used widely, and remain relatively cheap. The IGIS will continue to be focused on customization and relatively expensive solutions, but the added value will remain that much higher. In this market place, niche players will continue to plough a fertile furrow.

The development of IGIS will not necessarily replace simpler, cheaper packaged systems, because the two address different market requirements. Nor do customized solutions have to be re-invented for each application: they can be built from a set of modular software and analysis tools. Perhaps this will become easier as technology advances and data availability continues to explode. However, technological developments are in themselves insufficient. We must continue to strive towards a better understanding of business processes and problems. Such processes are often complex, and we must match that complexity in our solutions. If we fail, we will miss out on a great opportunity for geographers to forge long-lasting links with business and service planners, and we will open the way for the market consultants to dominate the market. It is such an opportunity that we simply cannot afford to throw it away.

APPENDIX: EXAMPLES OF PERFORMANCE INDICATORS IN RELATION TO TRAINING

RESIDENTIAL INDICATORS (AVAILABLE FOR ALL TARGET GROUPS)

1 Potential demand for training. This is simply O_i

(For this pilot study we assume all demand is actually met. If we could obtain greater information on the real flow patterns to training providers, then we could more accurately calculate percentage of demand that is actually met. This would also then allow us to calculate market penetration by different types of training providers.)

2 Distance travelled to obtain training:

$$\sum_j S_{ij} c_{ij} / \sum_j S_{ij}$$

(If most of the training is provided centrally then suburban areas will clearly have greater distances to travel.)

3 'Key problem areas': that is, a combination of high demand (at least five instances) and long travelling distances (>3 km). This is a 1 or 0 index.

4 Demand ratio: this is the level of potential demand per thousand of the population (of the appropriate target group).

5 Accessibility index (Hansen):

$$\sum_j W_j / c_{ij}$$

(measure of how close a residential or workplace location is to the training providers).

6 Measure of 'effective delivery': that is, how well are the target group served by training:

$$\sum_j (S_{ij} / S_{*j}) W_j \qquad (* \text{ denotes summation})$$

(similar to indicator 5, yet takes account of the expected interaction flows to training providers rather than simply proximity to all providers).

7 Provision ratio: this is effective delivery per head of demand.

FACILITY-BASED (TRAINING) INDICATORS

1 Number of courses provided;

   (basic location of training institutions).

2 Average distance travelled to centres:

   $$\sum_i S_{ij}\, c_{ij} / \sum_i S_{ij}$$

   (This is the opposite to residential indicators 2, that is, how far people are having to travel to the training providers.)

3 Catchment population:

   $$\sum_i \left(S_{ij}\, O_i / \sum_j S_{ij} \right)$$

   (This calculates the catchment area on the basis of how far people are actually travelling to use facilities, rather than simply using the circular drive time catchment areas common in the literature.)

4 Number of places: (W_j) split into target categories.

5 'Index of satisfaction': number of places available per head of catchment population: that is, indicator 4/indicator 3.

6 Catchment unemployment ratio: weighted average of residential indicator 4 across the catchment.

7 Catchment provision indicator: weighted average of residential indicator 7 across the catchment.

BIBLIOGRAPHY

Aangeenbrug, R.T. (1991) A critique of GIS. In D.J. Maguire, M.F. Goodchild, and D.W. Rhind (eds) *Geographical Information Systems: principles and applications.* Longman, London, vol. 1, 101–7.

Anselin, L. and Getis, A. (1993) Spatial statistical analysis and geographical information systems. In M.M.Fischer and P.Nijkamp (eds) *Geographical Information Systems, Spatial Modelling and Policy Evaluation.* Springer-Verlag, Berlin, 35–50.

Applebaum, W. (1965) Can store location be a science? *Economic Geography,* 41, 234–7.

Armstrong, D., Britten, N., and Grace, J. (1988) Measures of General Practitioner referrals: patient workload and list size effects. *Journal of the Royal College of General Practioners,* 38, 494–97

Ashworth, G. (1990) *Selling the City: marketing approaches in public sector planning.* Belhaven, London.

Aspinall, R. and Boxer, A. (1993) City of Wakefield MDC – land charges implementation. In Proceedings of the 'GIS 93' Conference, Birmingham, May. Blenheim, London, 157–70.

Badillo, A.S. (1993) Transportation and navigation. In G.H. Castle III (ed.) *Profiting from a Geographical Information System.* GIS World Inc., Fort Collins, 161–76.

Bailey, T.C. (1992) Statistical spatial analysis and geographic information systems: a review of the potential and progress in the state of the art. In Proceedings of the Third European Conference on Geographical Information Systems, EGIS 92, Munich, March, 186–203.

Bailly, A.S., Coffey, W.J., Paelinck, J.H.P., and Polese, M. (1992) *Spatial Econometrics of Services.* Avebury, Aldershot.

Barnekov, T., Boyle, R., and Rich, D. (1989) *Privatism and Urban Policy in Britain and the United States.* Oxford University Press, Oxford.

Batchelor, R.A. (1975) Household technology and the domestic demand for water. *Land Economics,* 3, 208–23.

Batey, P. and Brown, P.J. (1995) From human ecology to customer targeting: the evolution of geodemographics. In P. Longley and G.P. Clarke (eds) *GIS for Business and Service Planning.* GeoInformation, Cambridge, 77–103.

Batty, M. (1989) Urban modelling and planning: reflections, retrodictions and prescriptions. In B. Macmillan (ed.) *Remodelling Geography.* Blackwell, Oxford, 147–69.

Batty, M. (1992) Urban modelling in computer-graphic and geographic information system environments. *Environment and Planning, B, Planning and Design,* 19, 663–88.

Batty, M. (1993) Using geographical information systems in urban planning and policy-making. In M.M. Fischer and P. Nijkamp (eds) *Geographical Information Systems, Spatial Modelling and Policy Evaluation.* Springer-Verlag, Berlin, 51–72.

Batty, M. and Longley, P. (1994) *Fractal Cities.* Academic Press, London.

Batty, M. and Mackie, S. (1972) The calibration of gravity, entropy and related models of spatial interaction. *Environment and Planning A,* 4, 205–33.

Batty, M. and Xie, Y. (1994a) Modelling inside GIS: part 1: model structures, exploratory data analysis and aggregation. *International Journal of GIS, 8(3),* 291–307.

Batty, M. and Xie, Y. (1994b) Modelling inside GIS: part 2: selecting and calibrating urban models using ARC/INFO. *International Journal of GIS, 8(5),* 451–70.

Beaumont, J.R. (1986) Modelling should be more relevant: some personal reflections, *Environment and Planning A, 18,* 419–21.

Beaumont, J.R. (1991a) GIS and market analysis. In D.J. Maguire, M.F Goodchild and D.W. Rhind (eds) *Geographical Information Systems: principles and applications.* Longman, London, vol. 2, 139–51.

Beaumont, J.R. (1991b) Managing information: getting to know your customers. *Mapping Awareness, 5(1),* 17–20.

Beaumont, J.R. (1991c) *An Introduction to Market Analysis,* CATMOG 53. Geo-Abstracts, Norwich.

Bennett, R.J. (1985) Quantification and relevance. In R. Johnston (ed.) *The Future of Geography,* Methuen, London, 211–24.

Bennison, D.J. and Davies, R.L. (1980) The impact of town centre shopping schemes in Britain: their impact on traditional retail environments. *Progress in Planning, 14,* 1–104.

Benoit, D. and Clarke, G.P. (1995) Assessing retail location planning using GIS. Working Paper, School of Geography, University of Leeds.

Berkeley, J.S. (1976) Reasons for referral to hospital. *Journal of the Royal College of General Practitioners, 26,* 28–30

Berkhout, F., Clarke, M., Forte, P., and Claydon, D. (1985) The implications of the GP referral system for health authority planning: some observations and the results of a survey. Working paper 407, School of Geography, University of Leeds.

Bertuglia, C.S., Clarke, G.P., and Wilson, A.G. (eds) (1994a) *Modelling the City: planning, performance and policy.* Routledge, London.

Bertuglia, C.S., Clarke, G.P., and Wilson, A.G. (1994b) Models and performance indicators in urban planning: the changing policy context. In C.S. Bertuglia, G.P. Clarke and A.G. Wilson (eds) *Modelling the City.* Routledge, London, 20–36.

Bevan, R.G. and Spencer, A.H. (1984) Models of resource policy of regional health authorities. In M. Clarke (ed.) *Planning and Analysis in Health Care Systems.* Pion, London, 90–118.

Birkin, M. (1993) Not so much geography: more a style of life. *Research Plus,* 9–10 December, 6.

Birkin, M. (1994) Understanding retail interaction patterns: the case of the missing performance indicators. In C.S. Bertuglia, G.P. Clarke, and A.G. Wilson (eds) *Modelling the City,* Routledge, London, 105–20.

Birkin, M. (1995) Customer targeting, geodemographics and lifestyle approaches. In P. Longley and G.P. Clarke (eds) *GIS for Business and Service Planning.* GeoInformation, Cambridge, 104–49.

Birkin, M. and Clarke, G.P. (1991) Spatial interaction in geography. *Geography Review, 4(5),* 16–24.

Birkin, M. and Clarke, G.P. (1995) Using microsimulation methods to synthesize census data. In S. Openshaw (ed.) *A Census Handbook.* Longman, London, 363–87.

Birkin, M., Clarke, G.P., Clarke, M., and Wilson, A.G. (1987) Geographical information systems and model-based locational analysis: a case of ships in the night or the beginnings of a relationship? Working Paper 497, School of Geography, University of Leeds.

Birkin, M., Clarke, G.P., Clarke, M., and Wilson, A.G. (1990) Elements of a model-based GIS for the evaluation of urban policy. In L. Worrall (ed.) *Geographical Information Systems: developments and applications*, Belhaven, London, 133–62.

Birkin, M., Clarke, G.P., Clarke, M., and Wilson, A.G. (1994) Application of performance indicators in urban modelling: subsystems framework. In C. Bertuglia, G.P. Clarke and A.G. Wilson (eds) *Modelling the City*. Routledge, London, 121–50.

Birkin, M. and Clarke, M. (1988) SYNTHESIS – a synthetic spatial information system for urban and regional analysis: methods and examples. *Environment and Planning A, 20,* 1645–71.

Birkin, M. and Clarke, M. (1989) The generation of individual and household incomes at the small area level. *Regional Studies, 23(6),* 535–48.

Birkin, M., Clarke, M., and George, F. (1995) The use of parallel computers to solve non-linear spatial optimisation problems: an application to network planning. *Environment and Planning A, 27,* 1049–68.

Birkin, M. and Wilson, A.G. (1986) Industrial location models: 1, a review and an integrating framework. *Environment and Planning A, 18,* 175–205.

Blakemore, M.J. (1991) Managing an operational GIS: the UK National On-line Manpower Information System (NOMIS). In D.J. Maguire, M.F. Goodchild, and D.W. Rhind (eds) *Geographical Information Systems: principles and applications.* Longman, London, vol. 1, 503–13.

Boalt, A. and Bernow, R. (1991) STOCAB – a strategic information system for the Stockholm Region: recent developments and applications. In L. Worrall (ed.) *Spatial Analysis and Spatial Policy using Geographical Information Systems.* Belhaven, London, 102–26.

Boddy, M. (1992) Evaluating local labour market policy: the case of the TECs. *British Journal of Education and Work, 5(3),* 43–56.

Boden, P., Stillwell, J.C.H., and Rees, P.H. (1992) Internal migration projection in England: the OPCS/DOE model. In J.C.H. Stillwell and P. Congdon (eds) *Migration Models: macro and micro approaches.* Belhaven, London, 262–86.

Bolton, R. (1985) Regional econometric models. *Journal of Regional Science, 25,* 495–520.

Bondi, L. and Domash, M. (1992) Other figures in other places: on feminism, postmodernism and geography. *Environment and Planning D, Society and Space, 10,* 199–213.

Bracken, I. (1991) A surface model of population for public resource allocation. *Mapping Awareness, 5(6),* 35–8.

Bradshaw, W.P. and Zalzala, Y. (1988) Proposals for out of town shopping centres in Greater Manchester. Discussion Paper 36, Department of Geography, University of Salford.

Brail, R.K. (1990) Integrating urban information systems and spatial models. *Environment and Planning B, 17(4),* 417–27.

Brassel, K.E. and Reif, D. (1979) A procedure to generate Thiessen Polygons. *Geographical Analysis, 11(3),* 288–303.

Breheny, M.J. (1987) The context for methods: the constraints of the policy process on the use of quantitative methods. *Environment and Planning A, 19,* 1449–62.

Breheny, M.J. (1988) Practical methods of retail location planning. In N. Wrigley (ed.) *Store Choice, Store Location and Market Analysis.* Routledge, London.

Bromley, R. and Coulson, M. (1989) The value of corporate GIS to local authorities. *Mapping Awareness, 3(5)*, 32–5.

Brown, P.J. (1991) Geodemographics: a review of recent developments and emerging issues. In I. Masser and M. Blakemore (eds) *Handling Geographic Information*. Longman, London, 221–58.

Brown, P.J., Hirschfield, A.F., and Batey, P.W. (1991) Applications of geodemographic methods in the analysis of health condition incidence data. *Papers in Regional Science, 70*, 329–44.

Brown, S. (1992) *Retail Location: a micro-scale perspective*. Avebury, Aldershot.

Bruton, M. and Nicholson, D. (1987) *Local Planning in Practice*. Hutchinson, London.

Bryan, N.S. (1995) 'Business geographics' – a US perspective. In P. Longley and G.P. Clarke (eds) *GIS for Business and Service Planning*. GeoInformation, Cambridge, 250–70.

Burrough, P.A. (1991) Soil information systems. In D.J Maguire, M.F. Goodchild, and D.W. Rhind (eds) *Geographical Information Systems: principles and applications*. Longman, London, vol. 2, 153–69.

Burton, D. (1994) *Financial Services and the Consumer*. Routledge, London.

Buttenfield, B.P. and Mackaness, W.A. (1991) Visualisation. In D.J. Maguire, M.F. Goodchild, and D.W. Rhind (eds) *Geographical Information Systems: principles and applications*. Longman, London, vol. 1, 427–43.

Buxton, R. (1989) Integrated spatial information systems in local government – is there a financial justification? *Mapping Awareness, 2(6)*, 14–16.

CACI (1993) CACI's Insite system in action. *Marketing Systems Today, 8(1)*, 10–13.

Campbell, H. (1994) How effective are GIS in practice? A case study of British local government. *International Journal of GIS, 8(3)*, 309–26.

Castle, G. (1993) *Profiting from a GIS*. GIS World Books, Fort Collins.

CCN (1993) EuroMOSAIC. Product brochure, CCN Marketing, Nottingham.

Cesario, F. (1975) Linear and non-linear regression models of spatial interaction. *Economic Geography, 51*, 69–77.

Chadwick, G. (1971) *A Systems View of Planning*. Pergamon, Oxford.

Champion, A.G., Green, A.E., Owen, D.W., Ellin, D.J., and Coombes, M.G. (1987) *Changing Places: Britain's demographic and social complexion*. Edward Arnold, London.

Champion, A.G. and Townsend, A.R. (1990) *Contemporary Britain: a geographical perspective*. Edward Arnold, London.

Church, R., Current, J., and Eiselt, H. (1993) Editorial. *Locational Science, 1*, 1–3.

Church, R. and Schoepfle, O. (1993) The choice alternative to school assignment. *Environment and Planning B, 20*, 447–57.

Clarke, D.B. and Bradford, M. (1989) The use of space by advertising agencies within the United Kingdom. *Geografiska Annaler B, 71*, 139–51.

Clarke, G.P. and Clarke, M. (1995) The development and benefits of customised spatial decision support systems. In P. Longley and G.P. Clarke (eds) *GIS for Business and Service Planning*. GeoInformation, Cambridge, 227–45.

Clarke, G.P., Kashti, A., McDonald, A.T., and Williamson, P. (1995) Estimating small-area household demand for water. Working Paper, School of Geography, University of Leeds.

Clarke, G.P., Longley, P., and Masser, I. (1995) Business, geography and academia in the UK. In P. Longley and G.P. Clarke (eds) *GIS for Business and Service Planning*. GeoInformation, Cambridge, 272–81.

Clarke, G.P. and Wilson, A.G. (1994a) Performance indicators in urban planning: the historical context. In C.S. Bertuglia, G.P. Clarke, and A.G. Wilson (eds) *Modelling the City*. Routledge, London, 4–19.

Clarke, G.P. and Wilson, A.G. (1994b) A new geography of performance indicators. In C.S. Bertuglia, G.P. Clarke, and A.G. Wilson (eds) *Modelling the City*. Routledge, London, 55–81.

Clarke, M. (ed.) (1984) *Planning and Analysis in Health Care Systems*. Pion, London.

Clarke, M. (1986) Demographic forecasting and household dynamics: a microsimulation approach. In R. Woods and P.H. Rees (eds) *Population Structures and Models*. Allen and Unwin, Hemel Hempstead, 245–72.

Clarke, M. (1990) Geographical information systems and model-based analysis: towards effective decision support systems. In H. Scholten and J.C.H. Stillwell (eds) *Geographical Information Systems for Urban and Regional Planning*. Kluwer, Dordrecht, 165–75.

Clarke, M. (1993) Going beyond GIS: the next step to getting business the answers it needs. *Corporate Real Estate Executive, 8(2)*, 36–8.

Clarke, M., Chesworth, J., Macgill, S., McDonald, A., Siu, Y., and Wilson, A.G. (1993) Intelligent, interactive and analysis-based GIS: principles and applications. In P.M. Mather (ed.) *Geographical Information Handling: research and applications*. Wiley, London, 325–37.

Clarke, M. and Holm, E. (1987) Micro-simulation methods in human geography and planning: a review and further extensions. *Geografiska Annaler, 69B*, 145–64.

Clarke, M. and Rees, P.H. (1989) A microsimulation model for the diffusion of HIV and AIDS in the United Kingdom. Working Paper, School of Geography, University of Leeds.

Clarke, M. and Spowage, M.E. (1984) Integrated models for public policy analysis: an example of the practical use of simulation models in health care planning. *Papers of the Regional Science Association, 55*, 25–46.

Clarke, M. and Wilson, A.G. (1983) Modelling for health services planning. Working Paper 360, School of Geography, University of Leeds.

Clarke, M. and Wilson, A.G. (1984) Models for health care planning: the case of the Piemonte Region. Working Paper 36, Istituto Recherche Economico Sociale del Piemonte, Turin.

Clarke, M. and Wilson, A.G. (1985a) The dynamics of urban spatial structures: the progress of a research programme. *Transactions, Institute of British Geographers, 10(4)*, 427–51.

Clarke, M. and Wilson, A.G. (1985b) A model-based approach to planning in the National Health Service. *Environment and Planning B, Planning and Design, 12*, 287–302.

Clarke, M. and Wilson, A.G. (1987) Towards an applicable human geography: some developments and observations. *Environment and Planning A, 19*, 1525–42.

Clifford, D. (1993) The Vauxhall dealer area analysis package. *Mapping Awareness, 7(5)*, 25–7.

Cloke, P., Philo, C., and Sadler, D. (1991) *Approaching Human Geography*. Paul Chapman, London.

Colombo, G. and Comboni, G. (1991) The car industry. In R. Calori and P. Lawrence (eds) *The Business of Europe: managing change*. Sage, London, 116–36.

Connon, T. (1993) Tesco's bid to get off the shelf. *Independent on Sunday*, Business Section, 19 September, 6–7.

Coppock, J.T. and Rhind, D.W. (1991) The history of GIS. In D.J. Maguire, M.F. Goodchild, and D.W. Rhind (eds) *Geographical Information Systems: principles and applications.* Longman, London, vol. 1, 21–43.

Corina, M. (1971) *Pile it High, Sell it Cheap: the authorised biography of Sir John Cohen, founder of Tesco.* Weidenfeld and Nicolson, London.

Couclelis, H. (1991) Geographically informed planning: requirements for planning relevant GIS. *Papers in Regional Science, 70,* 9–20.

Cowen, D.J., Vang, A.H., and Waddell, J.M. Jr. (1983) Beyond hardware and software: implementing a state-level geographical information system. In E. Teicholz and B.J.L. Berry (eds) *Computer Graphics and Environmental Planning.* Prentice-Hall, Englewood Cliffs, NJ, 30–51.

Cox, R.K. (1968) *Retail Site Assessment.* Business Books, London.

Cresswell, P. (1995) Customised and proprietary GIS: past, present and future. In P. Longley and G.P. Clarke (eds) *GIS for Business and Service Planning.* GeoInformation, Cambridge, 192–226.

Crombie, D.L. and Flemming, D.M. (1988) General practitioner referrals to hospital: the financial implications of variability. *Health Trends, 20,* 53–6.

Cross, A. (1991) Who said GIS cannot fail? *Mapping Awareness, 5(10),* 12–14.

Dale, P.F. (1991) Land information systems. In D.J. Maguire, M.F. Goodchild, and D.W. Rhind (eds) (1991) *Geographical Information Systems: principles and applications.* Longman, London, vol. 2, 85–99.

Dangermond, J. (1983) Selecting new town sites in the United States using regional databases. In E. Teicholz and B.J.L. Berry (eds) *Computer Graphics and Environmental Planning.* Prentice-Hall, Englewood Cliffs, NJ, 119–40.

Daniels, P.W. (1985) *Service Industries: a geographical appraisal.* Methuen, London.

Danielson, L.E. (1979) An analysis of residential demand for water using micro time series. *Water Resources Research, 15(4),* 763–67.

Davies, R.L. (1977) Store location and store assessment research: the integration of some new and traditional techniques. *Transactions of the Institute of British Geographers,* 141–57.

Davies, R.L. (1984) *Retail and Commercial Planning.* Croom Helm, London.

Davies, R.L. and Howard, E.B. (1989) The Metro Centre experience and the prospects for Meadowhall. *Retail and Distribution Management, 17(3),* 8–12.

Davies, R.L and Rogers, D.S. (1984) *Store Location and Store Assessment Research.* Wiley, Chichester.

Dawson, J.A. and Broadbridge, A.M. (1988) *Retailing in Scotland 2005.* Institute for Retail Studies, Stirling.

Dawson, J.A. and Kirby, D.A. (1977) Problems and policies affecting the small shop. *International Journal of Physical Distribution, 7,* 244–54.

Dawson, J.A. and Kirby, D.A. (1979) *Small Scale Retailing in the UK.* Saxon House, Farnborough.

Day, P. and Klein, R. (1991) Variations in budgets of fund holding practices. *British Medical Journal, 303,* 168–70.

Densham, P.J. (1991) Spatial decision support systems. In D.J. Maguire, M.F. Goodchild, and D.W. Rhind (eds) *Geographical Information Systems: principles and applications.* Longman, London, vol. 1, 403–12.

Densham, P.J. and Rushton, G. (1988) Decision support systems for locational planning. In R. Golledge and H. Timmermans (eds) *Behavioural Modelling in Geography and Planning.* Croom Helm, London.

Department of Health (1989) *Working for Patients*. HMSO, London.

Department of Health (1992) *The Health of the Nation: a strategy for health in England*. HMSO, London.

Department of the Environment (1977) Development control policy note 13: large new stores. HMSO, London.

Department of the Environment (1987) *Handling Geographic Information*. Report to the Secretary of State for the Environment of the Committee of Enquiry into the Handling of Geographic Information (Chorley Report). HMSO, London.

Dibb, S. and Simkin, L. (1991) Targeting, segments and positioning. *International Journal of Retail Distribution and Management, 19(3)*, 4–10.

Dickinson, H.J. and Calkins, H.W. (1988) The economic evaluation of a GIS. *International Journal of GIS, 2(4)*, 307–27.

Ding, Y. and Fotheringham, A.S. (1992) The integration of spatial analysis and GIS. *Computers, Environment and Urban Systems, 16*, 3–19.

Domencich, T. and McFadden, D. (1975) *Urban Travel Demand: a behavioural analysis*. North-Holland, Amsterdam.

Duley, C.J., Rees, P.H., and Clarke, M. (1988) A microsimulation model for updating households in small areas between Censuses. Working Paper 515, School of Geography, University of Leeds.

Dzurik, A.A. (1990) *Water Resources Planning*, Rowman and Littlefield, Maryland.

Elkins, P. (1990) Water under the bridge – the background to Southern Water's experience with GIS. *Mapping Awareness, 4(8)*, 38–40.

Elliott, C. (1991) Store planning with GIS. In J. Cadoux-Hudson and D.I. Heywood (eds) *Geographic Information 1991*. The Yearbook of the Association for Geographic Information. Taylor and Francis, London, 169–72.

Epstein, B. (1971) Geography and the business of retail site evaluation and selection. *Economic Geography, 47*, 192–9.

Evans, R. (1992) Strategic development issues facing TECs in the North West. *British Journal of Education and Work, 5(3)*, 17–42.

Eyles, J. (1987) *The Geography of the National Health: an essay in welfare geography*. Croom Helm, London.

Eyles, J. and Donovan, J. (1990) *The Social Effects of Health Policy*. Avebury, Aldershot.

Fenwick, I. (1978) *Techniques in Store Location Research: a review and applications*. Retail and Planning Associates, Corbridge.

Fernie, J. (ed.) (1990) *Retail Distribution Management: a strategic guide to developments and trends*. Kogan Page, London.

Fienberg, S.E. (1970) An iterative procedure for estimation in contingency tables. *Annals of Mathematical Statistics, 41*, 907–17.

Finch, J. (1992) Spatial data and GIS at the Institute of Hydrology. *Mapping Awareness, 6(3)*, 17–20.

Flowerdew, R. (1991a) Spatial analysis in GIS. *Mapping Awareness, 5(6)*, 11–12.

Flowerdew, R. (1991b) Clustered residential area profiles and beyond. *Mapping Awareness, 5(3)*, 34–9.

Flowerdew, R. and Aitken, M. (1982) A method of fitting the gravity model based on the Poisson distribution. *Journal of Regional Science, 22*, 191–202.

Foot, D. (1981) *Operational Urban Models*. Methuen, London.

Fotheringham, A.S. (1983) A new set of spatial interaction models: the theory of competing destinations. *Environment and Planning A, 15*, 15–36.

Fotheringham, A.S. (1986) Modelling hierarchical destination choice. *Environment and Planning A, 18,* 401–18.

Fotheringham, A.S. (1992) Exploratory spatial data analysis and GIS. *Environment and Planning A, 24,* 1675–8.

Fotheringham, A.S. (1994) GIS and exploratory spatial data analysis, *Geographical Systems, 1(4),* 315–27.

Fotheringham A.S. and Rogerson, P. (eds) (1994) *Spatial Analysis and GIS.* Taylor and Francis, London.

Fotheringham, A.S. and Xie, Y. (1995) GIS-based analysis: a glimpse into the future of spatial policy formulation? Paper presented to the GIS for Public Policy Analysis RRL/ESRC Conference, N. Ireland, May.

French, S. and Wiggins, L. (1991) California planning agency experiences with automated mapping and geographical information systems. In L. Worrall (ed.) *Spatial Analysis and Spatial Policy using Geographical Information Systems.* Belhaven, London, 62–74.

Gardiner, V. and Herrington, P. (1986) *Water Demand Forecasting.* GeoBooks, Norwich.

Garner, B. (1990) GIS for urban and regional planning and analysis in Australia. In L.Worrall (ed.) *Geographical Information Systems: developments and applications.* Belhaven, London, 41–64.

Gatrell, A.C. (1988) Handling geographic information for health studies. Northern RRL Research Report 15, University of Lancaster.

Gatrell, A.C. and Vincent, P. (1990) Managing natural and technical hazards: the role of GIS. Regional Research Laboratory Initiative Discussion Paper 7, ESRC RRL, Department of Town and Regional Planning, Sheffield.

Gault, I. and Davis, S. (1988) The potential for GIS in a large urban authority: the Birmingham City Council Corporate GIS pilot. *Mapping Awareness, 2(5),* 38–41.

Gault, I. and Peutherer, D. (1990) Developing GIS in local government in the UK: case studies from Birmingham City Council and Strathclyde Regional Council. In L. Worrall (ed.) *Geographic Information Systems: developments and applications,* Belhaven, London, 109–32.

Ghosh, A. and Craig, C.S. (1984) A location-allocation model for facility planning in competitive environments. *Geographical Analysis, 16,* 39–56.

Ghosh, A. and Harche, F. (1993) Location-allocation models in the private sector: progress, problems and prospects. *Locational Science, 1,* 81–106.

Ghosh, A. and McLafferty, S.L. (1987) *Location Strategies for Retail and Service Firms.* Lexington Books, Lexington, MA.

Giggs, J. (1970) Socially disorganised areas in Barry: a multivariate approach. In H. Carter and W.K. Davies (eds) *Urban Essays: studies in the geography of Wales.* Longman, London, 101–43.

Goodchild, M. (1987) A spatial analytical perspective on GIS. *International Journal of GIS, 1(4),* 327–34.

Goodchild, M., Haining, R., and Wise, S. (1992) Integrating GIS and spatial data analysis: problems and possibilities. *International Journal of GIS, 6(5),* 407–23.

Gould, P. (1985) *The Geographer at Work.* Routledge, London.

Green, A. and Owen, D. (1992) Skill shortages and recruitment difficulties: data sources at a local level. *British Journal of Education and Work, 5(3),* 57–78.

Griffiths, D.A. (1993) Which spatial statistics techniques should be converted to GIS functions? In M.M. Fischer and P. Nijkamp (eds) *Geographical Information Systems, spatial modelling and policy evaluation.* Springer-Verlag, Berlin, 103–14.

Grima, A.P. (1971) *Residential Water Demand*. University of Toronto Press, Toronto.

Grimshaw, D.J. (1992) The use of GIS by building societies in the UK. Proceedings of the Third European Conference on GIS, Munich, vol. 2, 988–97.

Grimshaw, D.J. (1994) *Bringing Geographical Information Systems into Business*. Geo-Information, Cambridge.

Guy, C.M. (1981) The use of models and simulation methods in hypermarket impact studies. Papers in Planning Research No. 27, Department of Town Planning, UWIST, Cardiff.

Guy, C.M. (1991) Spatial interaction modelling in retail planning practice: the need for robust statistical methods. *Environment and Planning B, Planning and Design, 18*, 191–203.

Haggett, P. (1965) *Locational Analysis in Human Geography*. Edward Arnold, London.

Haining, R. (1993) *Spatial Data Analysis in the Social and Environmental Sciences*. Cambridge University Press, Cambridge.

Hale, D. (1991) The healthcare industry and geographic information systems. *Mapping Awareness, 5(8)*, 36–9.

Harris, B. (1989) Beyond geographical information systems: computers and the planning profession. *Journal of the American Planning Association, 55*, 85–90.

Harris, B. (1991a) Planning technologies and planning theories. Paper presented to the Joint Conference of the Association of European Schools of Planning and the Association of Collegiate Schools of Planning, Oxford, July.

Harris, B. (1991b) Planning theory and the design of planning support systems. Paper presented to the conference on Computers in Planning, Oxford, July.

Harris, B. and Wilson, A.G. (1978) Equilibrium values and dynamics of attractiveness terms in production-constrained spatial interaction models. *Environment and Planning A, 10*, 371–88.

Harrison, S., Hunter, D., Johnston, I., and Wistow, G. (1989) *Competing for Health: a commentary on the NHS review*. Nuffield Institute for Health Service Studies, University of Leeds.

Harrison, S. and Wistow, G. (1991) The purchaser/provider split in health care: towards explicit rationing. Paper presented to the British Section of the Regional Science Association, Oxford, September.

Hart, M., Bond, D., and Devine, P. (1991) Local labour market analysis in urban areas: an application of GIS technology in Northern Ireland. In L.Worrall (ed.), *Spatial Analysis and Spatial Policy using Geographical Information Systems*. Belhaven, London, 188–206.

Harvey, D. (1973) *Social Justice and the City*. Edward Arnold, London.

Haughton, G. and Peck, J. (1988) Skills audits: a framework for local economic development. *Local economy, 3(1)*, 11–19.

Haughton, G. and Peck, J. (1989) Local labour market analysis, skill shortages and the skills audit approach. *Regional Studies, 23(3)*, 271–6.

Haynes, K.E. and Fotheringham, A.S. (1984) *Gravity and Spatial Interaction Models*. Sage, Beverly Hills, CA.

Haynes, R. (1987) *The Geography of Health Services in Britain*. Croom Helm, London.

Hays, S.M., Kearns, R.A., and Moran, W. (1990) Spatial patterns of attendance at general practitioner services. *Social Science and Medicine, 31(7)*, 773–81.

Healey, A. and Ryan, M. (1992) A theoretical and empirical examination of the general practitioner's decision to refer: a preliminary attempt at explaining variations in referral practice. Discussion Paper 05/92, HERU, University of Aberdeen.

Heijt, P. (1994) Geographical information for retail marketing. Proceedings, GIS in Business '94 Europe. Longman, London, 66–8.

Higgs, B. and Wright, A. (1991) The 1991 Census and beyond – a GIS approach to demographic analysis in Dudley. *Mapping Awareness, 5(5)*, 38–42.

Hinton, M.A. and Wheeler, K. (1992) GIS in 1992 – towards a single financial European market. *Mapping Awareness, 6(1)*, 19–22.

Hirschfield, A., Brown, P., and Marsden, J. (1991) Database development for decision support and policy evaluation. In L. Worrall (ed.) *Spatial Analysis and Spatial Policy using Geographical Information Systems.* Belhaven, London, 152–87.

Hirschfield, A., Brown, P., and Marsden, J. (1993) Information systems for policy evaluation: a prototype GIS for urban programme impact appraisal in St Helens, North West England. In M. Fischer and P. Nijkamp (eds) *GIS, Spatial Modelling and Policy Evaluation.* Springer-Verlag, Berlin, 213–34.

HMSO (1956) *Report of the Committee of Inquiry into the cost of the NHS* (Guillebaud Committee). London.

HMSO (1980) *Report of the Working Group on Inequalities in Health* (Black Report). London.

HMSO (1984) *NHS Management Report* (Griffiths Report). London.

HMSO (1992) *Report of the Inquiry into London's Health Service, Medical Education and Research* (Tomlinson Report). London.

Hodgson, M.J. (1978) Toward more realistic allocation in location-allocation models: an interaction approach. *Environment and Planning A, 10*, 1273–86.

Howard, E. B. and Davies, R. L. (1990) The future of Meadowhall Shopping Centre. Research Paper D6. Oxford Institute of Retail Management, Templeton College, Oxford.

Howe, A. (1991) Assessing potential of branch outlets using GIS. In J. Cadoux-Hudson and D.I. Heywood (eds) *Geographic Information 1991.* The Yearbook of the Association for Geographic Information. Taylor and Francis, London, 173–5.

Howes, D. and Gatrell, A.C. (1993) Visibility analysis in GIS: issues in the environmental impact assessment of windfarm developments. Proceedings of the GIS 93 Conference, Birmingham, May. Blenheim, London, 245–60.

Huff, D.L. (1963) A probabilistic analysis of shopping center trade areas. *Land Economics, 39*, 81–90.

Huxhold, W. (1991) *An Introduction to Urban Geographical Information Systems.* Oxford University Press, New York.

IJRDM (1991) Issue on franchising, *International Journal of Retail Distribution and Management, 19(2)*, 4–19.

Ireland, P. (1994) GIS: another sword for St Michael. *Mapping Awareness*, April, 26–9.

Janssen, F. (1994) A portfolio approach for site location. Proceedings, GIS in Business '94 Europe. Longman, London, 223.

Jarman, B. (1983) Identification of underprivileged areas. *British Medical Journal, 286*, 1705–9.

Jefferys, J.B. (1954) *Retail Trading in Britain: 1850–1950.* Cambridge University Press, Cambridge.

Jenkins, J. and Walker, J.R. (1985) School roll forecasting. In J. England, K. Hudson, R. Masters, K. Powell, and J. Shortridge (eds) *Information Systems for Policy Planning in Local Government.* Longman, London, 96–112.

Johnson, N. (1990) *Reconstructing the Welfare State: a decade of change 1980–1990.* Harvester Wheatsheaf, London.

Johnston, R.J. (1991) *Geography and Geographers*. Edward Arnold, London.

Jones, K. and Simmons, J. (1990) *The Retail Environment*. Routledge, London.

Kay, W. (1987) *Battle for the High Street*. Piatkus, London.

Kearns, R.A. and Barnett, J.R. (1992) Enter the supermarket: entrepreneurial medical practice in New Zealand. *Environment and Planning C, Government and Policy, 10*, 267–81.

Kehris, E. (1990) A geographical modelling environment built around ARC/INFO. Proceedings of the First European Conference on Geographical Information Systems, EGIS Foundation, Utrecht.

Kim, K.E. (1990) The development and application of GIS in Hawaii. In L. Worrall (eds) *Geographical Information Systems: developments and applications*. Belhaven, London, 87–108.

King, N. and Newton, P. (1990) Patterns in practice. *Health Service Journal*, 3 May, 669.

Kingdom, B. (1991) The use of GIS for rehabilitation planning within the water industry. *Mapping Awareness, 5(1)*, 9–11.

Knox, P.L. (1978) The intra-urban ecology of primary medical care: patterns of accessibility and their policy implications. *Environment and Planning A, 10*, 415–36.

Knox, P.L. (1979) The accessibility of primary care to urban patients: a geographical analysis. *Journal of the Royal College of General Practitioners, 29*, 160–8.

Knudsen, D.C. and Fotheringham, A.S. (1986) Matrix comparison, goodness-of-fit, and spatial interaction modelling. *International Regional Science Review, 10(2)*, 127–47.

Kohsaka, H. (1993) A monitoring and locational decision support system for retail activity. *Environment and Planning A, 25*, 197–211.

Lake, R.W. (1993) Planning and applied geography: positivism, ethics and geographical information systems. *Progress in Human Geography, 17(3)*, 404–13.

Landis, J.D. (1993) GIS capabilities, uses and organisational issues. In G.H. Castle III (ed.) *Profiting from a Geographical Information System*. GIS World Inc., Fort Collins, 23–54.

Langston, P., Clarke, D., and Clarke, G.P. (1995) Retail saturation, retail location and retail competition: an analysis of British grocery retailing. Working Paper, School of Geography, University of Leeds.

Leathard, A. (1990) *Health Care Provision: past, present and future*. Chapman and Hall, London.

Lee, D.B. (1973) Requiem for large scale models. *Journal of the American Institute of Planners, 39*, 163–8.

Leonardi, G. (1983) The use of random utility theory in building location-allocation models. In J.F. Thisse and H.G. Zoller (eds) *Locational Analysis of Public Facilities*. North-Holland, Amsterdam, 357–83.

Leung, Y. (1993) Towards the development of an intelligent decision support system. In M.M. Fischer and P. Nijkamp (eds) *Geographical Information Systems, Spatial Modelling and Policy Evaluation*. Springer-Verlag, Berlin, 131–46.

LGMB (1991) *Geographic Information Steering Group: summary report of consultancy project*. Local Government Management Board, Luton.

Loney, M. (1983) *Community against Government: the British Community Development Project*. Heinemann, London.

Longley, P. and Clarke, G.P. (eds) (1995) *GIS for Business and Service Planning*. Geo-Information, Cambridge.

Lopez, X.R. and John, S.A. (1993) Data exchange and integration in UK local government. *Mapping Awareness, 7(5)*, 37–40.

Lord, J. and Lynds, C. (1981) The use of regression models in store location research. *Akron Business and Economic Review, 10,* 13–19.

Lovett, A. and Gatrell, A.C. (1988) The geography of spina bifida in England and Wales. *Transactions, Institute of British Geographers, 13(3),* 288–302.

Lowry, I.S. (1964) *A Model of Metropolis.* Rand, New York.

Lucas, H.C. Jr. (1978) *Information Systems Concepts for Management.* McGraw-Hill, New York.

McCullagh, M.J. and Ross, C.G. (1980) Delaunay Triangulation of a random data set for isarithmic mapping. *Cartographic Journal, 17(2),* 93–9.

McDonald, A.T., Clarke, M., Bramley, E., and Freestone, R. (1993) CATNAP and PWQIS: spatial tools for river and potable water quality evaluation. Proceedings of the IFIP TC5/WG5.11 Working Conference on Computer Support for Environmental Impact Assessment, Como, October.

McDonald, A.T. and Kay, D. (1987) *Water Resources: issues and strategies.* Longman, London.

McGoldrick, P.J. (1988) Spatial price differentiation by chain store retailers. In E. Kaynak (ed.) *Transnational Retailing.* Walter de Gruyter, Berlin, 167–80.

McGoldrick, P.J. (1990) *Retail Marketing.* McGraw-Hill, London.

McLoughlin, J.B. (1973) *Control and Urban Planning.* Faber, London.

McManners, P. (1992) Military survey and GIS. *Mapping Awareness, 6(3),* 65–7.

Maguire, D.J. (1995) Implementing spatial analysis and GIS applications for business and service planning. In P. Longley and G.P. Clarke (eds) *GIS for Business and Service Planning.* GeoInformation, Cambridge, 171–91.

Maguire, D.J., Goodchild, M.F., and Rhind, D.W. (eds) (1991) *Geographical information systems: principles and applications.* Longman, London.

Mahoney, R. (1989) Should local authorities use a corporate or departmental GIS? *Mapping Awareness, 3(2),* 57–9.

Mahoney, R.P. (1991) GIS and the utilities. In D.J. Maguire, M.F. Goodchild, and D.W. Rhind (eds) *Geographical Information Systems: principles and applications.* Longman, London, vol. 2, 101–14.

Marinker, M., Wilkin, D., and Metcalfe, D.H. (1988) Referral to hospital: can we do better? *British Medical Journal, 297,* 461–4.

Mark, D.M. (1987) Recursive algorithm for determination of Proximal (Thiessen) Polygons in any metric space. *Geographical Analysis, 19(3),* 264–71.

Markham, R., Rix, D., and Bentley, R. (1990) GIS at Plymouth City Council – coming to terms with reality. *Mapping Awareness, 4(8),* 10–13.

Marmot, M.G., Adelstein, A.M., and Bulusu, L. (1984) *Immigrant Mortality in England and Wales, 1970–1978: causes of death by country of birth.* HMSO, London.

Marsh, C., Arber, S., Wrigley, N., Rhind, D., and Bulmer, M. (1988) The view of academic social scientists on the 1991 UK Census of Population: a report of the Economic and Social Research Council Working Group. *Environment and Planning A, 20,* 851–5.

Martin, D. (1991) *GIS and their Socio-economic Applications.* Longman, Harlow.

Martin, D. (1995) Censuses and the modelling of population in GIS. In P. Longley and G.P. Clarke (eds) *GIS for Business and Service Planning.* GeoInformation, Cambridge, 48-72.

Martin, D., Longley, P., and Higgs, G. (1992) The geographical incidence of local government revenues: an intra-urban case study. *Environment and Planning C, 10,* 253–65.

Martin, D. and Williams, H.C.W.L. (1992) Market area analysis and accessibility to primary health care centres. *Environment and Planning A, 24*, 1009–19.

Mason, J.B. and Meyer, M.L. (1981) *Modern Retailing: theory and practice.* Business Publications, Plano, TX, 549.

Mason, R.U. and Matriff, I.I. (1981) *Challenging Strategic Planning Assumptions: theory, cases and techniques.* Wiley, New York.

Masser, I. and Blakemore, M. (1991) *Handling Geographical Information: methodology and potential applications.* Longman, Harlow.

Mayhew, L. (1982) A theory of health care facility location in cities: some notes. *Sistemi Urbani, 3*, 415–33.

Mayhew, L. (1986) *Urban Hospital Location.* Allen and Unwin, London.

Miller, E.J. and Lerman, S.R. (1981) Disaggregate modelling and decisions of retail firms: a case study of clothing retailers. *Environment and Planning A, 13*, 729–46.

Mohan, J. (1990) Spatial implications of the National Health Service White Paper. *Regional Studies 24*, 553–8.

Mohan, J. (1995) *A National Health Service? The restructuring of health care in Britain since 1979.* Macmillan, Basingstoke.

Monopolies and Mergers Commission (1992) *New Cars: a report on the supply of new motor cars within the United Kingdom.* HMSO, London.

Moore, A.T. and Roland, M.O. (1989) How much variation in referral rates among general practitioners is due to chance? *British Medical Journal, 298*, 500–2.

Moore, S. and Attewell, G. (1991) To be and where not to be: the Tesco approach to locational analysis. *Operational Research Insight, 4(1)*, 21–4.

Morrell, D.C., Gage, H.G., and Robinson, N.A. (1971) Referral to hospital by general practitioners. *Journal of the Royal College of General Practitioners, 21*, 77–85.

Mounsey, H.M. (1991) Multisource, multinational environmental GIS: lessons learnt from CORINE. In D.J. Maguire, M.F. Goodchild, and D.W. Rhind (eds) *Geographical Information Systems: principles and applications.* Longman, London, vol. 2, 185–200.

Mullen, P. (1990) Planning and internal markets. In P. Spurgeon (ed.) *The Changing Face of the NHS in the 1990s.* Longman, London, 18–37.

MVA/ULIS (1987) *Strategic Plan Review for East Yorkshire Health Authority.* East Yorkshire Health Authority, Hull.

Newkirk, R. (1987) A general purpose (generic) urban/regional modelling framework. *Environments, 19(1)*, 3–16.

Newkirk, R. (1991) Mapping metropolitan area futures: a case study from Toronto. In L. Worrall (ed.) *Spatial Analysis and Spatial Policy using Geographical Information Systems.* Belhaven, London, 207–31.

Newson, M. (1992) *Land, Water and Development.* Routledge, London.

Ody, P. (1989) Getting to know your customers. *Retail and Distribution Management*, July/August, 32–5.

Openshaw, S. (1986a) *Nuclear Power: siting and safety.* Routledge, London.

Openshaw, S. (1986b) Modelling relevance. *Environment and Planning A, 18*, 143–7.

Openshaw, S. (1989a) Computer modelling in human geography. In B. Macmillan (ed.) *Remodelling Geography.* Blackwell, Oxford, 70–88.

Openshaw, S. (1989b) Making geodemographics more sophisticated. *Journal of the Market Research Society, 31*, 111–31.

Openshaw, S. (1991a) A view on the GIS crisis in geography or, using GIS to put Humpty-Dumpty back together again. *Environment and Planning A, 23(5)*, 621–28.

Openshaw, S. (1991b) Developing appropriate spatial analysis methods for GIS. In D.J. Maguire, M.F. Goodchild, and D.W. Rhind (eds) *Geographical Information Systems: principles and applications*. Longman, London, vol. 1, 389–402.

Openshaw, S. (1992a) Further thoughts on geography and GIS: a reply. *Environment and Planning A, 24(4)*, 463–66.

Openshaw, S. (1992b) Some suggestions concerning the development of AI tools for spatial analysis and modelling in GIS. *Annals of Regional Science, 26*, 35–51.

Openshaw, S. (1993) GIS 'crime' and GIS 'criminality'. *Environment and Planning A, 25*, 451–8.

Openshaw, S., Charlton, M., Wymer, C., and Craft, A.W. (1987) A mark I geographical analysis machine for the automated analysis of point data sets. *International Journal of Geographic Information Systems, 1*, 335–58.

Openshaw, S. and Craft, A. (1991) Using geographical analysis machines to search for evidence of clusters and clustering in childhood leukaemias and non-Hodgkin lymphomas in Britain. In G. Draper (ed.) *The Geographical Epidemiology of Childhood Leukaemias and Non-Hodgkin Lymphomas in Britain, 1966–1983*. Studies in Medical and Population Subjects No. 53. HMSO, London, 109–26.

Openshaw, S., Wymer, C., and Charlton, M. (1986) A geographical information and mapping system for the BBC Domesday optical disks. *Transactions, Institute of British Geographers, 11(3)*, 296–304.

Ottens, H. (1990) The application of geographical information systems in urban and regional planning. In H. Scholten and J.C.H. Stillwell (eds) *Geographic Information Systems for Urban and Regional Planning*. Kluwer, Dordrecht, 15–22.

Pacione, M. (1974) Measures of the attraction factor. *Area, 6*, 279–82.

Parrott, R. and Stutz, F.P. (1991) Urban GIS applications. In D.J. Maguire, M.F. Goodchild, and D.W. Rhind (eds) *Geographical Information Systems: principles and applications*. Longman, London, vol. 2, 247–60.

Peel, R. (1993) Traffic director for London. *Mapping Awareness, 7(3)*, 8–10.

Penny, N. and Broom, D. (1988) The Tesco approach to store location modelling. In N. Wrigley (ed.) *Store Choice and Store Location*. Routledge, London, 109–19.

Phillips, D.R. (1979) Spatial variations in attendance at general practitioners' services. *Social Science and Medicine, 130*, 169–76.

Pickles, J. (1991) Geography, GIS and the surveillant society. *Papers and Proceedings of Applied Geography Conferences, 14*, 80–91.

Pickles, J. (ed.) (1995) *Ground Truth: the social implications of geographical information systems*. Guildford Press, New York.

Pinch, S. (1985) *Cities and Services: the geography of collective consumption*, Routledge, London.

Raper, J., Rhind, D., and Shepherd, J. (1992) *Postcodes: the new geography*. Longman, London.

RAWP (1976) *Sharing Resources for Health Care in England*. Report of the Resource Allocation Working Party. DHSS, London.

Rector, J.M. (1993) Utilities. In G.H. Castle III (ed.) *Profiting from a Geographical Information System*. GIS World Inc., Fort Collins, 193–208.

Rees, G. and Rees, P.H. (1991) *A Ward Population Information System for Swansea City Council*. GMAP, Leeds.

Rees, P.H., Clarke, M., and Duley, C.J. (1987) A model for updating individual and household populations. Working Paper 486, School of Geography, University of Leeds.

Rees, P.H. and Wilson, A.G. (1977) *Spatial Population Analysis*. Edward Arnold, London.

Revelle, C., Marks, D., and Liebman, J. (1970) An analysis of private and public sector location models. *Management Science, 16,* 692–9.

Reynolds, J. (1991) GIS for competitive advantage: the UK retail sector. *Mapping Awareness, 5(1),* 33–6.

Reynolds, J. (1993) The role of GIS in European cross-border retailing. *Mapping Awareness, 7(2),* 20–5.

Rhind, D.W., Goodchild, M.F., and Maguire, D.J. (1991) Epilogue. In D.J. Maguire, M.F. Goodchild, and D.W. Rhind (eds) *Geographical Information Systems: principles and applications*. Longman, London, vol. 2, 313–27.

Rhynsburger, D. (1973) Analytic delineation of Thiessen Polygons. *Geographical Analysis, 5(2),* 133–44.

Richard, D., Beguin, H., and Peeters, D. (1990) The location of fire stations in a rural environment. *Environment and Planning A, 22,* 39–52.

Robson, B.T. (1988) *Those Inner Cities: reconciling the economic and social aims of urban policy*. Clarendon Press, Oxford.

Rogers, D.S. and Green, H.L. (1979) A new perspective on forecasting store sales: applying statistical models and techniques in the analog approach. *Geographical Review, 69,* 449–58.

Roland, M.O. (1988) General practitioner referral rates: interpretation is difficult. *British Medical Journal, 297,* 437–8.

Roland, M.O., Bartholomew, J., Morrell, D.C., McDermott, A., and Paul, E. (1990) Understanding hospital referral rates: a user's guide. *British Medical Journal, 301,* 98–103.

Roy, G.G. and Snickars, F. (1993) Computer-aided regional planning – applications for the Perth and Helsinki regions. In M.M. Fischer and P. Nijkamp (eds) *Geographical Information Systems, Spatial Modelling and Policy Evaluation*. Springer-Verlag, Berlin, 181–98.

Rushton, G. (1988) Location theory, location-allocation models and service development planning in the Third World. *Economic Geography, 64,* 97–120.

Russac, D.A.V., Rushton, K.R., and Simpson, R.J. (1991) Insight into domestic demand from a metering trial. *Journal of the Institute of Water and Environmental Management*, June, 342–51.

Sayer, R.A. (1976) *A Critique of Urban Modelling*. Pergamon Press, Oxford.

Sayer, R.A. (1979) Understanding models versus understanding cities. *Environment and Planning A, 11,* 853–62.

Schiller, R. and Jarrett, A. (1985) A ranking of shopping centres using multiple branch numbers. *Land Development Studies, 2,* 53–100.

Scholten, H. and Padding, P. (1991) Working with GIS in a policy environment. *Environment and Planning B,* 405–16.

Scholten, H. and Stillwell, J.C.H. (eds) (1990) *Geographic Information Systems for Urban and Regional Planning*. Kluwer, Dordrecht.

Scott, A. (1971) *Combinatorial Programming: spatial analysis and planning*. Methuen, London.

Senior, M.L. (1991) Deprivation payments to GPs: not what the doctor ordered. *Environment and Planning C, 9,* 79–94.

Siderelis, K.C. (1991) Land resource information systems. In D.J. Maguire, M.F. Goodchild, and D.W. Rhind (eds) *Geographical Information Systems: principles and applications*. Longman, London, vol. 2, 261–73.

Simkin, L.P. (1990) Evaluating a store location. *International Journal of Retail and Distribution Management, 18(4),* 33–8.

Sleight, P. (1993) Targeting customers: how to use geodemographic and lifestyle data in your business. Press release, 15 October. NTC Publications, London.

Smith, D.A. and Tomlinson, R.F. (1992) Assessing costs and benefits of GIS: methodological and implementation issues. *International Journal of GIS, 6(3),* 247–56.

Smith, N. (1992) History and philosophy of geography: real wars, theory wars. *Progress in Human Geography, 16,* 257–71.

Smyth, F.M. and Thomas, R.W. (1994) Controlling HIV-AIDS in Ireland: the implications for health policy of some epidemic forecasts. SPA Working Paper 26, School of Geography, University of Manchester.

Sparks, L. (1986) The changing structure of distribution in retail companies: an example from the grocery trade. *Transaction, Institute of British Geographers, 11,* 147–54.

Spencer, A.H. (1978) Deriving measures of attractiveness for shopping centres. *Regional Studies, 12,* 713–26.

Spurgeon, P. (ed., 1990) *The Changing Face of the National Health Service in the 1990s.* Longman, London.

Starr, L.E. and Anderson, K.E. (1991) A USGS perspective on GIS. In D.J. Maguire, M.F. Goodchild, and D.W. Rhind (eds) *Geographical Information Systems: principles and applications.* Longman, London, vol. 2, 11–22.

Stillwell, J.C.H. (1994) Monitoring intercensal migration in the United Kingdom. *Environment and Planning A, 26(11),* 1711–30.

Streit, U. and Wiesmann, K. (1993) Problems of integrating GIS and hydrological models. Paper presented to the European Science Foundation Social Science Programme on GIS and Spatial Analysis, Amsterdam, December.

Swainston, R.A. (1993) GIS in local government: GIS at the crossroads. Proceedings of the GIS 93 Conference, Birmingham, May. Blenheim, London, 57–66.

Tactics International (1993) Tactician. Product brochure, Tactics International, London.

Taylor, P.J. and Johnston, R.J. (1995) Geographical information systems and geography. In J. Pickles (ed.) *Ground Truth: the social implications of geographical information systems.* Guildford Press, New York, 51–67.

Taylor, P. J. and Overton, M. (1991) Further thoughts on geography and GIS. *Environment and Planning A, 23(8),* 1087–94.

Thomas, R.W. (1992) *Geomedical Systems: intervention and control.* Routledge, London.

Thomasson, M. (1989) GIS in the National Rivers Authority. *Mapping Awareness, 3(5),* 37–40.

Thorpe, Bernard, and Partners (1988) *Comparison Goods Retailing in Leeds.* Bernard Thorpe and Partners, London.

Timmermans, H.J. (1981) Multi-attribute shopping models and ridge regression analysis. *Environment and Planning A, 13,* 43–56.

Tong, K.M. (1989) Resource allocation and utilisation at the small area level: results from the analysis of the Leeds Health Districts. Unpublished MA thesis, School of Geography, University of Leeds.

Townsend, P. and Davidson, N. (1982) *Inequalities in Health.* Penguin, Harmondsworth.

Townsend, P., Phillimore, P., and Beattie, A. (1988) *Health and Deprivation: inequality and the North.* Croom Helm, London.

Towsey, R. (1972) Finding the right site. *Marketing,* June, 40–3.

Walters, D. and White, D. (1987) *Retail Marketing Management.* Macmillan, Basingstoke.

Ward, A. (1994) Saving lives in the West Country – using GIS to improve ambulance response times. *Mapping Awareness,* April, 36–7.

Webber, R.J. (1992) Streets ahead of the rest. *Precision Marketing,* 7 December.

Weibel, R. and Heller, M. (1991) Digital terrain modelling. In D.J. Maguire, M.F. Goodchild, and D.W. Rhind (eds) *Geographical Information Systems: principles and applications.* Longman, London, vol. 1, 269–97.

Wellar, B. (1993) GIS fundamentals. In G.H. Castle III (ed.) *Profiting from a Geographical Information System.* GIS World Inc., Fort Collins, 3–22.

West, A. (ed.) (1988) *Handbook of Retailing.* Gower, Aldershot.

Whitehead, M. (1987) *The Health Divide: inequalities in health in the 1980s.* Health Education Council, London.

Whitton, M. (1993) Optimising the use of a geographical information system: a case study of the GMAP RADAR system. Unpublished MA thesis, School of Geography, University of Leeds.

Wijkel, D. (1986) Lower referral rates for integrated health centres in the Netherlands. *Health Policy, 6,* 185–98.

Wilkin, D., Metcalfe, D.H., and Marinker, M. (1989) The meaning of information on GP referral rates to hospitals. *Community Medicine, 11,* 60–5.

Wilkin, D. and Smith, A.G. (1987) Explaining variation in general practitioner referral rates to hospital. *Family Practice, 4(3),* 160–9.

Williams, H.C.W.L. (1981) Random utility theory and probabilistic choice models. In A.G. Wilson, J. Coelho, S.M. Macgill, and H.C.W.L. Williams (eds) *Optimisation in Location and Transport Analysis.* Wiley, Chichester, 46–84.

Williams, P.A. and Fotheringham, A.S. (1984) The calibration of spatial interaction models by maximum likelihood estimation with program SIMODEL. Geographic Monograph Series, 7, Department of Geography, Indiana University.

Wilson, A.G. (1967) A statistical theory of spatial distribution models. *Transportation Research, 1,* 253–69.

Wilson, A.G. (1970) *Entropy in Urban and Regional Modelling.* Pion, London.

Wilson, A.G. (1974) *Urban and Regional Models in Geography and Planning.* Wiley, Chichester.

Wilson, A.G. (1983) A generalised and unified approach to the modelling of service supply structures. Working Paper 352, School of Geography, University of Leeds.

Wilson, A.G. (1989) Classics, modelling and critical theory: human geography as structured pluralism. In B. Macmillan (ed.) *Remodelling Geography.* Blackwell, Oxford, 61–9.

Wilson, A.G. and Bennett, R.J. (1985) *Mathematical Methods in Human Geography and Planning,* Wiley, Chichester.

Wilson, A.G. and Birkin, M. (1987) Models of agricultural location in a spatial interaction framework. *Geographical Analysis, 18,* 31–56.

Wilson, A.G., Coelho, J.D., MacGill, S.M., and Williams, H.C.W.L. (eds) (1981) *Optimisation in Location and Transport Analysis.* Wiley, Chichester.

Womack, J.P., Jones, D.T., and Roos, D.T. (1990) *The Machine that Changed the World.* Rawson, New York.

Wood, S.J. (1990) Geographic information system development in Tacoma. In H. Scholten and J.C.H. Stillwell (eds) *Geographic Information Systems for Urban and Regional Planning.* Kluwer, Dordrecht, 77–94.

Woodhead, K. (1985) Population projections. In J. England, K. Hudson, R. Masters, K. Powell, and J. Shortridge (eds) *Information Systems for Policy Planning in Local Government*. Longman, London, 76–95.

Worrall, L. (1989) Design issues for planning-orientated spatial information systems. *Mapping Awareness, 3(3)*, 17–20.

Worrall, L. (ed.) (1990a) *Geographic Information Systems: developments and applications*. Belhaven, London.

Worrall, L. (1990b) Information systems for urban and regional planning in the United Kingdom: a review. *Environment and Planning B, 17(4)*, 451–62.

Worrall, L. and Rao, L. (1991) The Telford Urban Policy Information Systems Project. In L. Worrall (ed.) *Spatial Analysis and Spatial Policy using Geographical Information Systems*. Belhaven, London, 127–51.

Worthington, S. (1995) The cashless society. *International Journal of Retail and Distribution Management, 23(7)*, 31–40.

Wrigley, N. (1985) *Categorical Data Analysis for Geographers and Environmental Scientists*. Longman, London.

Wrigley, N. (1993) Retail concentration and the internationalization of British grocery retailing. In R.D.F. Bromley and C.J. Thomas (eds) *Retail Change: contemporary issues*. UCL Press, London, 41–68.

Wrigley, N. and Lowe, M. (eds) (1995) *Retailing, Capital and Consumption: towards the new retail geography*. Longman, London.

Wrigley, N., Morgan, K., and Martin, D. (1988) Geographical information systems and healthcare: the Avon project. *ESRC Newsletter, 63*, 8–11.

Yeh, A.G. (1990) A land information system for the monitoring of land supply in the urban development of Hong Kong. In L. Worrall (ed.) *Geographical Information Systems: developments and applications*. Belhaven, London, 163–88.

SUBJECT INDEX

AUTHOR INDEX